The Secret Life of
GERMS

The Secret Life of
GERMS

Observations and Lessons from a Microbe Hunter

Philip M. Tierno, Jr., Ph.D.

POCKET BOOKS

New York   London   Toronto   Sydney   Singapore

 POCKET BOOKS, a division of Simon & Schuster, Inc.
1230 Avenue of the Americas, New York, NY 10020

Copyright © 2001 Philip M. Tierno, Jr., Ph.D.

ISBN: 0-7434-2187-6

First Pocket Books hardcover printing November 2001

10 9 8 7 6 5 4 3 2 1

POCKET and colophon are registered trademarks of Simon & Schuster, Inc.

For information regarding special discounts for bulk purchases, please contact Simon & Schuster Special Sales at 1-800-456-6798 or business@simonandschuster.com

Designed by Nancy Singer

Printed in the U.S.A.

This book is dedicated to

James E. Cimino, M.D.,

Michael J. Brescia, M.D.,

and Guenther Stotzky, Ph.D.

They showed me the way.

Acknowledgments

Walter Cronkite said it best when he wrote in his 1996 book, *A Reporter's Life,* that "acknowledgment isn't nearly a strong enough word to express the gratitude I owe to all those who made this book possible." I wholeheartedly agree. Although I have written numerous scientific articles, including chapters in medical textbooks, I had only a vague idea about the process of writing a book for the general public. No one individual who writes such a book can engage in this activity alone. Where do I begin to thank the myriad individuals who have helped me make this work possible?

Let me first thank my professional mentors, Drs. James E. Cimino and Michael J. Brescia of Calvary Hospital in the Bronx, who showed me how to pursue a scientific career while simultaneously reaping enjoyment from life. They have infused in me a great appreciation and deep reverence for the gift of life. Their example set the high professional and ethical standards that have become my benchmark. I have been moved and irrevocably altered by the persuasive force of these gentlemen. They truly inspired me with joie de vivre!

Dr. Guenther Stotzky, the former chairman of biological sciences at New York University, my scientific mentor, instilled in me an appreciation of the entire environment in which microbes live. He taught me to weigh all of the physicochemical factors that help determine the ecological habitats of germs in order to discern the dynamics of their interactions with both man and the biosphere. But for him my understanding of microbial ecology would be incomplete. Dr. Edward Bottone of Mt. Sinai Medical Center in New York encouraged me to pursue a doctoral degree, and so helped launch me on a scientist's path. My thanks to them both.

The fertile academic atmosphere provided by New York University, New York University Medical Center, and the New York University School of Medicine has enabled me to prosper and grow during my thirty plus years of affiliation with this great institution. I am proud to say that New York University Medical Center was one of the first major health care centers in the United States to recognize the importance of environmental medicine. In 1947, NYU established a division of industrial medicine within its medical school which has become the Nelson Institute of Environmental Medicine at NYU Medical Center, and which has grown into the largest and most productive university-based environmental health research institute in the world. The interrelatedness of man, microbes, and the environment is so profoundly important that our very survival depends on recognizing and appreciating it. Over the years, the Nelson Institute has been at the forefront of defining that relationship.

Special thanks to Lynn Odell, Senior Director of Public Affairs for NYU Medical Center, and her staff, for providing me with an endless number of opportunities to speak to the media and thereby educate the public on germ matters of concern. I thank Dr. Mark Lifshitz, Director of Clinical Laboratories at NYU Hospitals Center for his friendship, collegiality, and outstanding leadership. Through the years, his words of encouragement and kindness have meant a great deal to me.

I thank the present and former chairmen of the department of microbiology at NYU School of Medicine, Drs. Claudio Basilico and Milton R. Salton, as well as the chairman of the department of pathology, Dr. Victorio De Fendi, for their faith in my ability. I thank Dr. Joel Oppenheim, Associate Dean for Graduate Studies, Sackler Institute of Graduate Biomedical Studies, NYU School of Medicine, for his valued support over the last twenty years of our professional and personal association.

Thanks to Kenneth Inglima, my treasured friend, colleague, and confidant, for the free exchange of ideas throughout our thirty years of friendship. It has been comforting to know he has always been there for me.

To my friend and colleague Dr. Bruce A. Hanna at New York University School of Medicine and Bellevue Hospital, many thanks for his advice and counsel over the twenty plus years of our friendship. We've journeyed together through the peaks and valleys of professional life and have weathered its storms. Through it all we and our families have maintained and enjoyed a precious personal relationship.

In working on this book I have had the benefit of four fine editors. I sincerely thank Elisa Petrini, who helped create a viable structure for the book. Her wisdom and insight provided a crucial impetus for getting the book under way. What can I say about Hilary Hinzmann, editor par excellence? His contribution was immeasurable. His great skills helped metamorphose this book into its final form. I am profoundly indebted to Hilary and am most appreciative of his extraordinary ability and his newfound friendship. And many thanks to Tracy Behar, editorial director of Pocket Books, and to Tracy Bernstein, senior editor at Pocket, for their insightful editing and for so adroitly and enthusiastically guiding the book to completion. Now comes Robert Tabian, my literary agent, who made these connections for me. The knowledge that he was "taking care of it," no matter whether the problem was large or small, was most reassuring during my most trying times. I owe him a great debt of gratitude.

Thanks to the Oprah Winfrey Show for planting the seed for the book by asking if any of the material that I was to present on a segment of the show was in print anywhere. About one year afterward I was in the green room before appearing on the Montel Williams Show, when Montel Williams suggested that I should "put this stuff in book form for people." He truly added fertilizer to the seed. But the water for the endeavor was added by a trenchmate from my formative years, Mary Ellen Ginnetti, RN. She added copious quantities of water, which enabled the seed to sprout. I am eternally indebted to Mary Ellen for her great contribution.

I am thankful for the contributions made by my secretary, Ann Sullivan; and my typist, Tina Sullivan, who completed the original drafts of my manuscript; and to Jenny Torres, for typing the proposal. Their patience and ability in deciphering my "chicken-scratch" are greatly appreciated. To Denise Saviano, my cyberinformant, thank you.

Finally, I could not have completed the overwhelming task of writing such a work without the cooperation of my wonderful family. Thanks to my wife and companion of thirty-four years, Josephine, who was my sounding board; and to my daughters, Alexandra and Meredith, and their husbands, François and Thomas, who were my most ardent critics. I thank my mother and father for their unfailing support throughout my life, and my mother-in-law and father-in-law for their never-wavering interest in my career.

The Germ

A mighty creature is the germ,
Though smaller than a pachyderm.
His customary dwelling place
Is deep inside the human race.
His childish pride he often pleases
By giving people strange diseases.
Do you, my poppet, feel infirm?
You probably contain a germ.

—Ogden Nash

Contents

Acknowledgments   vii

Part I

The Animalcules   1

  1   Seeds of Disease, Seeds of Life   3
  2   How We Make Each Other Sick   10
  3   The Germ Domain   30
  4   The Germ Factory   42
  5   The Enemy Within   62

Part II

The Seeds of Disease   87

  6   Person to Person   89
  7   Common Ground   118
  8   In Thin Air   149
  9   On the Wing and on Eight Little Legs   170

Part III

The New Age   199

  10   Not the Usual Suspects   201
  11   The Germ Revolt   222

Contents

12 Germ Warfare and Terrorism 236

13 Germs for Life 251

Index 279

PART I

The Animalcules

The number of these animalcules in the
scurf of a man's teeth are so many, that
I believe they exceed the number of
men in a kingdom.

—Antoni van Leeuwenhoek, 1684

Seeds of Disease, Seeds of Life

Our Greatest Fear

Think of Howard Hughes, and what comes to mind? Is it a picture of a handsome, vital man, heir to a great fortune, whose exploits as an aviator, a movie producer, a husband of starlets, a wily businessman, and a Nixon campaign contributor have filled scores of articles, books, and movies about him? Or does another image predominate, the lonely figure of an aging, unkempt, drug-addicted recluse, ensconced at vast expense in a Las Vegas hotel suite that he obsessively tries to keep operating-room clean, fretting endlessly that some germ will infect his system and kill him?

The picture of the germ-phobic old man that Howard Hughes became lingers in the public imagination because, when it comes to germs, we all have a little bit of Howard Hughes in us. We're all infected with the psychic fear that at any moment, in any setting, invisible agents may as easily give us an incurable, lethal disease as they would a common cold. Or perhaps they will poison our food, or attack our children as they play with their friends and pets. We know that nasty germs could be anywhere, and so we feel helpless to predict when they'll strike or to prevent the harm they can do to us, those we love, and our communities.

There is just cause for alarm. Except for the very earliest stages of human history, infectious diseases have been, and remain, the number-one killer worldwide. Thanks to advanced medical care and public sanitation, infectious diseases run a close third to heart disease, the number-one killer, and cancers, the number-two killer, in the United States and other developed countries. But that calculation may need to be revised. As we'll see, the latest scientific evidence implicates infectious germs as a trigger for many cases of heart disease and many kinds of cancer.

Once upon a time, not too long ago, we thought we had defeated germs. In the middle of the twentieth century, the development of "wonder drugs" promised to end the scourge of infectious disease forever. Beginning with penicillin, scientists and medical researchers soon stocked our pharmacies with a wide range of antibiotics and vaccines. It seemed to be the culmination of one of the great themes in the saga of human history, the long struggle to understand the underlying causes of disease. As we'll see, over the millennia people in many different cultures achieved profound insights into the nature and progress of disease—the ancient Chinese, for example, invented a dangerous but effective method of inoculating against smallpox—but the means to put all the pieces of the puzzle together were lacking. Superstition, greed, egotism, and inadequate technology all played roles in keeping us ignorant. Not until the Renaissance period in Western Europe did a critical mass of knowledge become available to scientists and physicians working in a new atmosphere of free inquiry.

These circumstances enabled an Italian physician named Girolamo Fracastoro to theorize that diseases were transmitted by tiny agents, too small to be seen by the naked eye, which he called "seminaria," that is, "seeds" of disease. Fracastoro published his book on contagious diseases, the first statement of the modern germ theory of disease, in 1546. Just over a hundred years later Antoni van Leeuwenhoek in Holland and Robert Hooke in England used the new optical technology of the microscope to demonstrate that "seminaria" actually existed. But Fracastoro's theory would not be proved in full until the late nineteenth century, when the Frenchman Louis Pasteur, the German Robert Koch, and others finally established reliable procedures for identifying specific disease-causing germs, mapping their transmission from host to host, and developing vaccines to combat them.

The great achievements of Pasteur and Koch set the stage for the twentieth century's discovery of penicillin and all the wonder drugs that have followed to this day, including the latest "cocktails" being administered to patients with HIV, the virus that causes AIDS. Unfortunately, as the AIDS epidemic has made all too clear, we can no longer confidently expect an easy triumph over infectious disease. Even as our wonder drugs and vaccines have allowed us to eliminate scourges such as smallpox, probably the greatest single killer in the history of the human species, we face a frightening array of newly emergent diseases and of old diseases in new guises. Killer germs that once fell easy prey to a few doses of antibiotics, such as those that cause tuberculosis, have reappeared in drug-resistant forms. Some germs defy all known antibiotics. Meanwhile, globalization has unleashed germs that were once confined among isolated population groups—HIV, Ebola virus, West Nile virus, among others—and made it possible for them to spread from anywhere to anywhere, just like a computer virus on the Internet, only much more deadly. In the context of global commerce, travel, tourism, and mass migration, there is no longer any such thing as an isolated population group. When it comes to public health, if nowhere else, we must acknowledge that the world's population is truly all one family.

In addition, medical research is continually uncovering new disease roles for germs, from toxic shock syndrome to cancer and heart disease. This is not even to mention the continuing threat of germ warfare or bioterrorism with lethal agents such as anthrax. Less than a thermosful of anthrax germs could kill every warm-blooded creature in, say, Chicago—people, cats, dogs, horses, and the rest—leaving the Miracle Mile a desolate province of insects, reptiles, and birds. That is, it could do so if a way could be found to spread it effectively. Fortunately for us, that is no easy task, but, as we'll explore, it is also not necessarily beyond the reach of a committed terrorist group.

Although we must ultimately look to medical science to save us from all these dangers, the medical profession itself has a dirty germ secret that can no longer be ignored. I mean literally dirty. Every year our very best hospitals sicken and even kill an untold number of patients with what are called nosocomial infections. That medical euphemism, "nosocomial," means that it was doctors, nurses, and other staff who made the patients sick, because these caregivers' hands were contaminated with germs. The source of the contamination? The caregivers' failure to wash their hands properly, or wash them at all, after examining another patient, handling a specimen, using medical equipment, or attending to their own personal hygiene.

How is it possible that doctors at elite teaching hospitals don't wash their hands enough? Don't they know any better? Of course they do. And every good hospital now has a hand-washing program in place. Next time you're in a hospital, you might look for some of the signs that remind medical personnel about hand washing. These signs often feature a picture of two clasped hands and the words "Wash me."

The real importance of the fact that doctors don't always wash their hands as they should is that it indicates the extent to which inadequate personal hygiene cuts across all socioeconomic lines. This is not just a problem of agricultural workers and restaurant staffs. When researchers put cameras in public rest rooms to track people's behavior, the numbers of those who don't wash their hands properly, or don't wash them at all, is staggering, often over ninety percent. The rates are such that we really need a vast public-education campaign on hand washing, equivalent to that we now have on smoking. If we look at the matter honestly, we'll see that the cost to public health from not washing our hands is enormous.

Why so few people wash their hands appropriately is baffling. One part of the answer may well be the complacency about infections that the ready availability of antibiotics has encouraged until recently. But it is hard not to think that the behavior is mainly fueled by ignorance or selfishness. From the point of view of public health, there is a clear link between not covering one's mouth when one coughs and not fully disclosing one's sexual history to a prospective partner. They are related behaviors, and the threat posed by them gets ever greater, as our world becomes ever more populous and more closely connected.

Our Best Hope

What can we do to protect ourselves, our families, and our communities? Well, we can't rid the world of germs. For one thing, we depend on a great many germs to keep our economy churning along. Germs are vital to numerous agricultural, commercial, and medical processes, from the yeasts that make bread rise, through the microscopic algae used in manufacturing cosmetics, paints, and fertilizers, to the soil bacteria from which antibiotics are extracted.

More to the point, the cycle of life requires the action of germs at every stage. No living creature could survive for long in an entirely germ-free environment. Without germs, animals, including human beings, could not develop mature immune systems or even digest their food (as germs break down food in the intestine, they extract and produce essential nutrients and vitamins). The ecosystem of the human body, if you will, is delicately balanced by germs. Often when we get sick, the problem is that we have disturbed this natural balance and turned our own good and necessary germs against us. We can see that there is a similar balance in the world as a whole, if we consider the role that germs play at the end of life. If there were no germs to decompose them, the dead carcasses of animals and plants would soon cover the earth, choking off all future growth. Recently we have learned to utilize this capacity of germs to break down organic matter to help clean up oil spills and other pollution.

At an even deeper level, we have discovered that germs were the initial building blocks of evolution. They are much more than just the seeds of disease. They are the seeds of life itself. A fossilized germ cell found in a rock in western Australia and dated to 3.5 billion years ago is the oldest known sign of life on earth. In the beginning was the germ. With all that time in which to multiply, mutate, and adapt to diverse conditions, it should not surprise us to learn that thousands upon thousands of species of germs have colonized every corner of the Earth, from the ocean floor and the frozen wastelands of Antarctica to boiling mineral hot springs and the lava cones of volcanoes. The air we breathe, the water we drink, the food we eat, the ground we walk on, the surfaces we touch, all of it is a teemingly populous, roiling sea of germs. Germs inhabit every inch of our skin, and every channel of our bodies. In fact, some germs and some exposure to germs throughout life are vital to human health and immunity. There are more germs in our intestines than there are stars in the sky, some thousand billion germs per gram of matter. The number of germ cells in a human body actually exceeds the number of body cells by a factor of ten. And the combined weight of microscopic germs exceeds the combined weight of all living animals and plants. Germs can survive, even thrive, at radiation counts a thousand times higher than the level that would kill a human being; they have been recovered, alive, from the petrified gut of a forty-million-year-old bee and from a 250-million-year-old piece of frozen brine.

Germs fill an important niche even when they die, as fossilized germs become a significant part of the surface structure of the Earth. The stones used to construct the pyramids in Egypt, for example, largely consist of the shells of countless protozoa, one of the many varieties of germs we'll take a look at in the course of this book.

Germs are so important in the ecology of the world that alien observers might conclude that they are the dominant life-form on our planet. These observers might logically see us human beings as just one of many species that have evolved from germs and continue to live among them. If we could have the observational data these aliens might compile, we could learn an enormous amount about our place in a "germ healthy" world.

That's actually a good working description of what human science has tried to do in studying germs. The continuing benefits of that effort can improve the quality of life of everyone on the planet. In the course of a lifetime's involvement in studying germs, including the experience, as I will recount later, of helping to solve the mystery of toxic shock syndrome, I have gained an enduring admiration for the astonishing variety of germs and their seemingly limitless adaptability. My fascination with germs began in childhood, when I read the life story of Louis Pasteur. Around the same time I saw on television a Hollywood movie about Pasteur, starring Paul Muni. Then I read Paul de Kruif's classic book, *The Microbe Hunters,* about Pasteur and the other great pioneers of microbiology, and that got me hooked. Right then I knew that I wanted to become a microbe hunter, too, and take part in the quest to solve the germ puzzles that lie at the heart of human health and disease, and indeed of all biology.

The more we discover of the secrets of germs, the most genetically resourceful and varied living things on our planet, the more we shall understand about the fundamental workings of biology and the better able we will be to fight cancer and other as yet incurable diseases. My work has also given me an abiding appreciation for humanity's ability, slowly but surely, to lift itself out of ignorance and adapt to change. We will learn the secrets of the germ—we are learning them now—and the pace of discovery, as we'll see, grows ever faster. In the meantime, the science of germs has already given us effective strategies we can employ now, at home, school, work, and elsewhere, to safeguard ourselves, our families, and our communities from infectious illness and deadly contagions.

Today we are truly standing at the dawn of a new age. As we begin a new millennium, we are finally uncovering the deepest biochemical and genetic workings of germs. In so doing, we are gaining the capacity to use this knowledge for the benefit of humanity. These smallest of creatures may one day allow us to resolve the most pressing human problems of disease, hunger, and pollution. We must continue to explore the gargantuan potential of germs so that we may harness their power for the good of mankind. How strange yet fitting that the future of nature's greatest creature, man, depends upon an intimate cooperation with nature's least, the germ!

2

How We Make Each Other Sick

It Could Happen to You

In the public rest rooms of Grand Central Terminal in midtown Manhattan, a scientific experiment is under way. The behavior of all the people who use the rest rooms is being observed without their knowledge. The researchers conducting the experiment want to find out the answer to a simple question: Who will, and who will not, wash their hands before leaving?

A busy train station is the perfect place for the experiment, because the variety of people passing through it insures that the researchers will get to observe a representative cross section of the population. In Grand Central, a major New York City subway station as well as an end point for travel in and out of the city, that cross section includes homeless people, commuters, suburbanites in town to shop or see a show, workers from station businesses, some of which sell food, and people from the surrounding neighborhood. People of every age, race, gender, ethnicity, and social class pass through the station every day, and as some of each group use the rest rooms, cameras record whether or not they wash their hands.

The results of the experiment don't surprise the researchers, but they are dramatic. The numbers confirm the results of similar experiments: well over sixty percent of people fail to wash their hands at all, and fewer than ten percent really wash their hands adequately. Almost no one cleans his or her hands thoroughly with soap and water and then avoids touching any surfaces before leaving the rest room. Men, women, and children, blue-collar and business-class, people generally don't wash their hands at all, or they only sprinkle a little water on their fingertips, as the high nobility did at the court of France's Louis XIV, the Sun King, because they feared the diseases that water itself might transmit.

As far as hygiene is concerned, that little sprinkle of water is no improvement over not washing. Both behaviors will leave microscopic germs on the hands, ready to be brought near the eyes, nose, and mouth, or an open cut or sore, all primary infection sites, as people unconsciously touch their faces and bodies, greet one another with handshakes and kisses, or exchange objects such as the food that is prepared and served in a restaurant.

Just consider this imaginary, but very possible, scenario. Three men stand side by side at the urinals in a rest room at Grand Central. To the left is a twenty-something associate in a law firm. When he got off his subway train from Brooklyn a few minutes before, he bought a bagel with cream cheese to eat at his desk; it's now in a paper bag stuffed into his messenger-style shoulder bag. To the right is a corporate insurance executive in his fifties, just off the train he takes every day from Nyack to his midtown office. In between them is a refugee expert for the United Nations, who's spent the past two weeks in Africa and Southeast Asia, gathering field assessments for a special report to the secretary-general. As soon as he reaches his office, he has to start finalizing his covering memorandum. He worked on it on the journey home, which began in Indonesia twenty-seven hours ago, but the beginning still doesn't feel quite right.

The last leg of the UN expert's journey was a red-eye flight from Los Angeles. After it landed in New York, he raced home to drop his bags and change into some clean clothes. He was just about to leave his apartment thirty minutes ago, when he became nauseated and threw up. Luckily, he didn't get anything on himself, as far as he could see; all he had to do was brush his teeth again. Then he went out the door, regretting that he'd eaten that yucky-looking breakfast on the plane that morning. You can't trust airline food, he thinks, especially when you've been moving through so many time zones in so few days. He's always had an irritable stomach, and unfortunately it seems to be getting more irritable with every passing year.

Unbeknownst to the man from the UN, however, he contracted Ebola virus about a week ago in Uganda. Now the virus is spotting his hands, which have microscopic particles of vomit on them. He will soon be very sick. Even as he stands at the urinal he starts to feel flushed and overheated. Sweat breaks out on his forehead and the palms of his hands. The virus teems in all his bodily fluids, in fact, including his urine, and that too is now spotting his hands.

Serendipitously, all three men finish at the urinals at the same time. As the UN expert steps away from his urinal, he feels as if he's going to faint and he stumbles a little. The twenty-something lawyer and the middle-aged insurance executive both reach out to grab him. They steady him before he falls, and for a moment he stands between them, gripping the lawyer's right hand in his left hand and the insurance executive's left hand in his right hand. As he does so, he transfers the Ebola virus to each through the vomit, urine, and sweat on his hands. To the naked eye, his hands still look clean.

"Whew! This is a little embarrassing," the UN expert says. "I ate some bad food a little while ago without realizing it, I think." He lets go of the Good Samaritans' hands, and they ask him if he's sure he's all right. He insists he is, thanks them again, and heads out the door.

The young lawyer follows right after him. Neither bothers to wash his hands. When the lawyer reaches his office, he goes straight to his desk to eat his bagel. In the process, he ingests the Ebola virus. In two weeks he will be as sick as the UN expert will become over the next couple of days. The prognosis for both is dire. As it will be for anyone else they infect.

As for the insurance executive, he's the son of a cleanliness fanatic. Whenever he hears the expression "You could eat off the floor in that house," he recalls his mother's pride in keeping things spotless and how much trouble he used to get into for tracking dirt inside. Then there were the nuns at parochial school, who conducted cleanliness checks every day after lunch and recess, and who whacked many a dirty hand with their rulers. They always hit with the metal edge, too. So now, a creature of habit, the insurance exec goes from the urinal to the sinks and carefully washes his hands. In the process, he sheds all traces of the Ebola virus before it has a chance to enter his system or to infect someone else he touches.

This scenario may be imaginary, but it is all too possible. It demonstrates that cleanliness in public rest rooms is not just a question of what people naturally touch inside the rest room, but also of whatever microscopic organisms they pick up elsewhere and then carry inside to deposit on surfaces. In so doing they leave residues of germs for others to touch and become contaminated with. Even a pristine-looking bathroom may provide fertile breeding grounds for germs.

What makes this important is the fact that eighty percent of all infectious illnesses, from the common cold to flesh-eating bacteria and lethal viruses like Ebola, are transmitted by touch. This happens either directly, by contact with another person, or indirectly, by contact with something that person has touched. In this light, hand washing emerges as a public-health issue that is every bit as serious as smoking, if not more serious. As I mentioned in the introduction, the issue extends to hospitals, where caregivers' dirty hands regularly produce unnecessary illnesses and deaths. As we'll see later, the infections people get in hospitals can be particularly nasty. The very fact that hospitals use large amounts of antibiotics makes them growth zones for dangerous antibiotic-resistant germs.

There are two and a half million of these nosocomial infections every year. And every year they directly cause thirty thousand deaths and contribute substantially to another seventy thousand. A hundred thousand deaths a year and an annual cost to society of $4.5 billion for treatment after the infections occur, when simple hand washing could prevent the overwhelming majority of them. Nosocomial infections kill more people every year than pancreatic cancer, leukemia, multiple sclerosis, Parkinson's disease, and Alzheimer's combined. These diseases are the subject of large public-relations campaigns to raise awareness and solicit funds to combat them. As yet there is nothing of the kind for the much more serious matter of illness and death from nosocomial infections.

The number of deaths from nosocomial infections, most of

them one hundred percent preventable, should make anyone stop and think before going into a hospital for any period, however brief, for any treatment or procedure, however minor. Even if you're only going to the hospital as an outpatient for the day or a few hours, you should ask, "What can I do to protect myself from the very real risk that the hospital will make me sicker than I was when I went in, maybe even kill me?" Or perhaps your partner or child has to go to the hospital. Later on I will discuss how to lower these risks, but the matter has clearly reached a crisis point and needs to be addressed in terms of society, and the medical profession, as a whole. More about that later, too.

Germ City

Of course, as the Ebola scenario suggests, you don't have to go into a hospital to become sick by touching someone or something. It's germ city, and indeed germ world, out there. Recently I taught a reporter working on an article about germs for a weekly newspaper, the *New York Observer,* how to collect bacterial samples from thirty-four suspected germ hot spots in New York, ranging from the backseat of a taxicab to the engagement-ring counter at Tiffany's and the rest rooms at the Waldorf Hotel. The reporter, Alex Kuczynski, and her associates took samples during the summer, when heat and high humidity create ideal conditions for germs to thrive. Sure enough, the samples contained a dazzling profusion of microbes, enough to warrant a front-page story headlined, "New York Is Germ City."

When I tested the sample taken from a taxicab seat, it contained *Streptococcus viridans,* a common mouth germ probably expelled by a cough; *Enterobacter* sp, found in feces; and *Staphylococcus aureus,* shed from skin. A movie-theater seat sample teemed with *S. aureus* as well as group B strep, which is usually found in vaginal fluid, and *Enterococcus,* another fecal germ. A woman wearing shorts or a short skirt likely deposited these germs on the seat. There were high counts of *S. aureus* and *Escherichia coli,* yet another fecal germ, on the headsets at a movie theater. Skin and fecal germs were also found on bar stools in various places and on the armrest of a chair in a corporate cafeteria. An apparently spotless glass at the bar in an exclusive Upper East Side hotel was actually spotted, invisibly, with group D strep, another of the many germs passed in feces; no doubt, a bartender or busboy handled the glass improperly. *Pseudomonas aeruginosa,* a common environmental organism which can produce a whopping eye infection, thrived on a nail dryer at an exclusive beauty salon and in the shower at a fashionable sports club. The sports-club shower also contained *Klebsiella pneumoniae;* as its name indicates, this germ, which is found naturally in the human intestine, can cause penumonia if it manages to get into the lungs. Finally, among other germ-ridden locales, consider a public phone at Madison Avenue and Seventy-first Street. Lingering on the handset were *S. aureus* and beta-hemolytic strep group A, a flesh-eating bacterium.

These are the sorts of germs we run into every day. Some of them, like *S. aureus* shed from skin or the intestinal bacteria that get passed in feces, are part of our bodies' normal complement of germs and are necessary for our health, as I'll explain later. But if enough of them get in or on the wrong part of the body, serious illness, even death, can result. For example, if *S. aureus* infects an open sore or cut and goes untreated, it could lead to a minor infection or possibly to something as potentially lethal as toxic shock syndrome. The relatively innocuous fecal germs found on seats, stools, and armrests could just as easily have been parasitic amoebae, Norwalk virus, or *Salmonella,* all of which can cause diarrhea. Fortunately, people with healthy immune systems usually beat off most germs pretty easily, including some of the nastier ones. What causes an illness to take root is not just contact with an infectious germ, but the kind and quantity of germs, the frequency or length of contact, and the health of the individual.

Some population groups are at extra risk of getting sick from infectious germs. Children, pregnant women, the elderly, and those with depressed immune systems may be vulnerable when others are not. The group B strep found on the movie-theater seat could have been a risk to a pregnant woman, for example, because it can cause neonatal meningitis. But the lack of adequate hand washing in our society can catch up with almost anyone sooner or later, including the minority who do wash their hands regularly and well (although they are better protected than the rest of the population, by far). If and when it does, count yourself lucky if all you come down with is a cold, a bout of diarrhea from food poisoning, or some other self-limiting infection. (A self-limiting infection is one that runs its course in a healthy person without doing any serious or lasting damage.)

The Grand Central Terminal rest room experiment, the *New York Observer* article, and many similar studies, formal and informal, demonstrate conclusively that, given the germy state of things, most of our society doesn't wash hands as frequently as is really advisable. It's as if people, including the doctors and nurses who don't wash their hands often enough, had no idea that invisible agents with the potential to sicken and kill were everywhere around them. One very worrying aspect of this is that children aren't learning better habits. Sadly, cameras in rest rooms reveal how seldom adults help children to wash their hands. One can only hope that there is more hand washing being done at home.

Despite the advances in knowledge that modern medical science has achieved, when it comes to germs it's out of sight, out of mind for most people. The only exception is the temporary public anxiety that breaks out when the media announce the emergence or reappearance of a killer germ, like those involved in AIDS, Ebola, or West Nile fever. The costs of getting sick—in time, money, and suffering—could be dramatically reduced, if only people would wash their hands more. But as history and common sense teach, it takes time, effort, money, and most of all understanding for individuals and societies alike to get and stay clean.

For much of human history that understanding has been fragmentary at best. It has taken millennia to acquire accurate knowledge of germs, their role in disease, and the benefits of cleanliness. Only in the last century and a half have scientists established the main facts of how germs work. So perhaps it shouldn't be surprising that habits like bathing and showering regularly are still very recent and not always practiced consistently, even in developed Western societies. Regular showering has been common in the Western world for about fifty years; people took to the practice only after most homes became centrally heated and there was an adequate supply of hot water.

In the developing world, enormous numbers of people still live as all of our ancestors used to do. Forty percent of the world's population—2.4 billion people—has no access to proper sanitation and clean water. As many as five million people, the majority of them children, die every year from acute diarrhea resulting from contaminated food and water, not to mention the toll of other germ diseases like tuberculosis. If the plight of people in developing countries does not give us pause on their account, it should on our own. In a global economy of mass migrations and frequent air travel, there is no such thing as an isolated population group, no guaranteeing that Third World germs will not invade First World communities.

Cleanliness Is Next to Godliness

When I showed the reporter from the *New York Observer* how to collect bacterial samples, there was one spot that seemed likely to host a multitude of germs, because of the large number of people who touch it every day. Yet the spot, a door handle at St. Patrick's Cathedral, was virtually germ-free. Hearing this, my secretary said, "God made sure there were no germs there."

The idea that "cleanliness is next to godliness," as John Wesley, the founder of Methodism, put it in the late eighteenth century, has a long history. In Old Testament times, the Hebrew civilization stressed cleanliness and made sanitary and dietary laws a part of religious observance. In many ways, the history of good hygiene is the history of understanding disease, and vice versa. The records of that history, beginning about 3000 B.C., fall into three rough periods: from 3000 B.C. to about A.D. 500, the time of the great civilizations of the ancient world; from A.D. 500 to A.D. 1500, the centuries of the Middle Ages; and A.D. 1500 to the present, when recognizably modern sciences established themselves and became engines of discovery.

I want to point out some of the important highlights of those periods, but the time before recorded history has a few things to teach us, as well. In every human society, there have always been people who served as healers. From the earliest times, shamans, witches, and medicine men and women have called on religious belief, magic, and the power of suggestion to heal and comfort the sick. But they have also used concrete knowledge gained through careful observation and passed down from generation to generation. Ancient healers could set fractures and perform surgeries, and they amassed impressive knowledge of effective treatments using herbs and other natural substances. They usually tried to restrict this information to themselves, because it was their stock-in-trade, and shamanic traditions kept some important facts of health a secret for centuries, if not millennia. In the late 1700s, for example, an English witch told a physician named William Withering the secret of digitalis, a medicine derived from the foxglove plant that is now commonly used in treating heart disease. Today pharmaceutical companies pay close attention to what researchers can glean from indigenous peoples in the Amazon and elsewhere about medicinal plants.

The knowledge that primitive peoples gained from observing the natural world and the course of illness in both human beings and other animals became the foundation for the slow growth of medical science. The most important idea of primitive medicine may not seem at all scientific at first. It's the notion that illness is the result of being out of balance in some way with the forces of the universe, whether the imbalance is imagined as being the work of gods, spirits, demons, or some mysterious natural cause. One way or another, humanity has kept pegging away at this idea of being in balance with one's environment, developing more sophisticated versions of it to keep pace with more accurate knowledge of how all living things—animals, plants, and germs—form one vast, complex ecosystem. The difference today is that instead of speculating about imbalances in climate, diet, temperament, or the spirit world, as shamans did, modern scientists and researchers can track the spread of specific germs from one location to another, in the world at large or in the delicately balanced ecosystem of the human body. You might say that today we use antibiotics to restore a healthy balance between germs and people, and vaccines to prevent an imbalance from occurring.

But that's getting ahead of the story. From the point of view of germs and disease, primitive people's observations and trial-and-error efforts at healing surely produced one other significant idea, the notion that clean and dry is generally healthier than dirty and wet. This idea surfaced over and over again in recorded history before scientists were finally able to establish that disease-causing germs exist, and that they thrive in filthy, humid conditions. In fact, human societies throughout history can be ranked according to the value they put on cleanliness. If modern Western society is stacked up against ancient civilizations in these terms, the present isn't always a clear winner over the past.

3000 B.C. to A.D. 500

The first record of a central community water supply and drainage dates to 3000 B.C. in Nippur, a city in the Sumerian empire, which was located in the region known as Mesopotamia, a part of modern-day Iraq. The first reference to soap, a recipe for making it, is inscribed on clay tablets dating to the same time and region. At this point people had no way of figuring out how soap and water work on dirt and germs at the molecular level, but they could certainly see the effects of cleanliness. For a long time soap was a luxury item, difficult to manufacture. It was not until the early nineteenth century that new industrial processes made it possible to make good-quality soap in large quantities.

The ancient Egyptians put a great emphasis on cleanliness, and they cautioned against eating meat from sick animals. They also believed that diseases were caused by parasites. But without the aid of microscopes, they imagined that creatures big enough to be seen with the naked eye, such as insects and worms, were to blame. The ancient Egyptians were apparently the first people to practice a form of germ warfare, using diseased corpses to contaminate the food and water supplies of their enemies.

The period from 3000 B.C. to A.D. 500 saw the rise of sophisticated medical traditions in India, China, and the Middle East and Mediterranean. Although there are many differences among them, the Indian system of Ayurvedic medicine, traditional Chinese medicine with herbs and acupuncture, and classical Greek and Roman medicine were all based on humoral theories. The humors were elements thought to be at work in the cosmos or in human bodies. Illness occurred when these elements were out of balance. Ayurvedic physicians classified seven body substances (bone, flesh, fat, blood, semen, marrow, and digestive juices) that they saw as the product of three humors (phlegm, bile, and wind). Traditional Chinese medicine described five elements (metal, wood, water, fire, and earth) as embodying the relationship of yin and yang (representing male and female forces). And the Greeks and Romans influenced the course of medicine in the West for centuries with their belief in four humors (yellow bile, black bile, phlegm, and blood).

Quaint as these theories may seem today (of course, traditional Indian and Chinese medicine continue to be used by millions of people, including increasing numbers of people in Western societies), they did not stop ancient physicians and scientists from recognizing a number of interesting factors in the course of disease. For example, the earliest medical records in China, Egypt, and Mesopotamia all describe using moldy and fermented substances to treat wounds. This was, in effect, a use of antibiotics, which occur naturally in these substances. The practice dates at least as far back as 1500 B.C., but it probably began much earlier, in the trial-and-error experiments of medicine men and women.

Around 2000 B.C., Sanskrit writings advised boiling water to purify it. In the fifth century B.C., the Greek physician Hippocrates, whose famous oath graduating physicians still take, observed that respiratory illness peaks in the winter, when people tend to stay inside in close proximity to one another, and that malaria peaks in the summer. But he did not take the step of theorizing that anything approximating germs was involved in these ailments. In the second century A.D., Galen, the personal physician to the Roman emperor Marcus Aurelius, asserted that all diseases had purely natural causes; he also prescribed washing with soap and water to help cure skin diseases. In the fifth century A.D., the Hindu physician Susruta noted connections, which he could not explain without the benefit of germ theory, between mosquitoes and malaria and between rats and plague. And as I mentioned in Chapter 1, the ancient Chinese devised a method of inoculating against smallpox, using pus taken from recovering patients. This technique reached Europe only in 1720. It was effective, but dangerous because of the virulence of the smallpox virus in the pus.

These observations and a great many others that remain scientifically valid were like so many scattered pieces to the puzzle of how germs cause disease. But wherever there were large cities and towns, astute observers couldn't help noticing that the density of population and the accumulation of all manner of waste and garbage made fertile ground for epidemics. In the Roman world this motivated the creation of a network of aqueducts, sewers, and public baths that was not equaled, much less surpassed, until the great urban public-health projects of late-nineteenth-century Europe and America.

A.D. 500 to A.D. 1500

As the Roman empire broke up, much of the understanding of the need for effective public sanitation was lost. The ruling principle of early Christian Europe was in many ways obedience to authority, especially that of the Bible but also that of figures like Galen. The hierarchy of the Church also stressed curing the ills of the soul rather than those of the body. This did not mean that science stopped cold in the Dark Ages, as they are sometimes still called. But along with demographic, economic, and other factors, a reliance on received authority as the ultimate guide did slow scientific progress. In terms of understanding germs and their role in disease, a key event during this period was the thirteenth-century English scientist Roger Bacon's assertion that invisible natural entities, not supernatural ones, were responsible for disease. This had an impact on later scientists, but when the Black Death, the bubonic plague, began to ravage Europe half a century later, beginning in 1348, most people saw it as a divine punishment for sins or attributed it to other imaginary causes.

During these centuries, medicine and science fared better in the Islamic world. Although Islamic scientists and physicians, like their Christian counterparts, looked to Galen and other ancient writers for guidance, they also discovered basic chemical processes such as distillation. The tenth-century Islamic physician Rhazes was able to recognize that smallpox and measles, which have some of the same symptoms, were actually different diseases. And thirteenth-century Cairo saw the establishment of the first hospitals to have separate wards dedicated to specific diseases. Keeping patients apart in this way reduced the incidence of multiple infections and speeded recovery.

A.D. 1500 to the Present

In Renaissance Europe the pieces to the puzzle of germs and disease finally began to come together in a secure way. As I discussed in Chapter 1, in 1546 the Italian physician Girolamo Fracastoro published a book on contagious diseases that argued that "seminaria," invisible "seeds," were the cause of disease. Fracastoro's concern with infectious diseases was heightened enormously by a contagious new venereal disease that was then sweeping across Europe. People variously named this disease the "French disease," the "Polish disease," and so on, blaming it on whichever of their neighbors they happened to despise and fear the most. Fracastoro keenly noted the onset and progress of the disease, and his observations influenced his theory of how all diseases spread. Italy and France were often bitter enemies at this time, and like other Italians Fracastoro thought of the new illness as the "French disease." But he also gave it the name we know it by today in a Latin poem he wrote called "Syphilis sive morbus gallicus," that is, "Syphilis or the French disease." The poem minutely describes the "foul sores" that break out on the body of a shepherd named Syphilis when he makes the mistake of angering the god Apollo.

Proof that "seminaria" such as those posited by Fracastoro actually existed came in 1676, when Antoni van Leeuwenhoek, a Dutch merchant, used a handheld lens to become the first person to see living germs. His detailed descriptions and simple drawings of what he called "animalcules," or "little animals," presented in letters to the Royal Society of London between 1676 and 1723, have withstood the test of time to the present day. His observations were so accurate that he was able to assert, rightly, that the "number of these animalcules in the scurf of a man's teeth are so many that I believe they exceed the number of men in a kingdom." In fact, there are more germs in the intestinal tract of a human being than the number of people who have ever lived.

People did not immediately put Leeuwenhoek's discoveries and Fracastoro's germ theory together, however. It would take two more centuries for that to happen. But after millennia of ignorance interrupted by fragmentary insights and logical conjectures, humanity at last knew of the existence of germs and could begin to trace their role in disease.

If science had been able to draw the connection sooner, the seventeenth and eighteenth centuries would probably have been a lot cleaner than they were. At the same time that Leeuwenhoek was establishing the existence of germs, the court of Louis XIV at Versailles was scrupulously avoiding washing too much. Physicians cautioned against daily bathing, because they believed that even clean water transmitted disease. They advised their patients to wash only the tips of the fingers, because they were used for eating, the cheeks, nose, and the area under the eyes, using a linen cloth called a "toilette," the origin of the word toilet. Apparently, the French physicians and their patients never noticed the logical contradiction between washing some areas of the body and not others. But they actually had some good logic on their side. At that time only the very well off had piped-in running water, and the pipes were made of wood, which naturally contained a great many microorganisms that would infect the water. So too much of that very impure but clean-looking water could indeed make a person sick.

In the eighteenth century, fashion changed and middle-class and upper-class people began to wash more of their bodies and to wash more frequently. But because hot water for bathing was difficult to provide, even in well-off households, an entire family would wash in the same tub of water. A sheet of fabric was placed on the bottom of a metal tub before the first person took a bath. The cloth served two functions. It protected the bather against being burned by the bottom of the tub. It also helped filter dirt and debris that would collect on the bottom of the tub as each person bathed. The sheet was taken out and replaced each time another person got into the bathwater.

The mass of people continued to live without washing much. In the first place they often didn't have access to clean water. And in the second, well, old habits die hard. Even in the nineteenth century, people bitterly resented the cold showers that armies and prisons began to insist on, and many physicians cautioned well-to-do patients that full immersion in water could cause disease.

After Leeuwenhoek, the next big breakthrough in understanding germs and disease came in 1796, when the English physician Edward Jenner discovered a safe vaccination for smallpox. Jenner noticed that people who had suffered from cowpox, a related but milder disease, were immune to smallpox, perhaps the greatest killer in human history. He inoculated an eight-year-old boy with material taken from cowpox pustules. After waiting a few weeks for the boy to fall ill with cowpox and recover, he inoculated him with material taken from smallpox pustules. This time the boy did not fall ill. It would take almost two hundred years for scientists to figure out how to manufacture sufficient quantities of a reliable vaccine and eliminate smallpox worldwide, but Jenner's inoculations represented the first significant step toward that goal.

Leeuwenhoek's research with a microscope and Jenner's experiment with inoculation proved conclusively that germs existed, that they caused disease, and that imitating their actions, as Jenner did with his vaccine, could combat disease. But no one knew where germs come from, or how ubiquitous they are in air, water, soil, and all living things. The consensus of scientific opinion was that germs arose by some kind of spontaneous generation on moldy or fermented materials. This was a continuation of an age-old theory that diseases resulted from miasmas, poisons spontaneously released into the air by stagnant water, swampy soils, and dead animals and plants. If miasmas made people sick, personal cleanliness couldn't matter very much. Those who thought the facts might be different were hard-pressed to make any headway against the establishment.

In the 1840s Ignaz Semmelweis, a Hungarian physician working in Vienna, did his level best just the same. A great problem of this time was puerperal, or childbed, fever, which often raged among poor women in hospital maternity wards. Death rates from childbed fever sometimes reached thirty percent. Better-off women usually gave birth at home, attended by a physician, midwife, or both. But if the poor wanted medical assistance during childbirth, they had to go to public hospitals.

Semmelweis noticed that medical students were going straight from the autopsy rooms to the maternity wards without washing their hands, and that the women they treated had very high rates of infection. He also knew that a colleague of his had died of fever after cutting himself during an autopsy. Although Semmelweis, like other scholars of his day, didn't realize that Leeuwenhoek's germs were the cause of disease, he theorized that the students were carrying a contagious agent of some kind from one place to another. He therefore instructed all the students working in his ward to wash their hands thoroughly with chlorinated lime before assisting at births. This resulted in an immediate, dramatic drop in deaths from childbed fever. Unfortunately the hospital authorities, outraged at Semmelweis's taking a policy matter into his own hands, so to speak, and refusing to accept his results as in any way conclusive, ordered the hand washing to stop.

In 1850, Semmelweis left Vienna and returned to Hungary, becoming professor of obstetrics at the University of Pest. There he again enforced a strict hand-washing regimen and again dramatically reduced deaths from childbed fever. But other doctors refused to adopt his practices. Semmelweis was haunted by the thought of so many women dying unnecessarily, and he was also deeply troubled when his two children died young from other causes. The combination apparently precipitated a mental breakdown, and Semmelweis died in a mental hospital in Vienna in 1865. In a tragic irony, the cause of death was an infected cut he had sustained while performing an autopsy.

Perhaps the most poignant aspect of Semmelweis's story is that he was a prophet only slightly before his time. In 1858, the Frenchman Louis Pasteur, one of the greatest medical scientists in history, finally put together the existence of germs, as established by Leeuwenhoek, and the germ theory of disease, as formulated by Fracastoro. That year he announced his first important research results in a paper titled, "A History of Lactic Acid Fermentation," showing that specific germs caused specific kinds of fermentation and stating that specific germs also caused many different diseases. As he wrote in a somewhat later paper, diseases resulted from "bodies . . . resembling in all points the germs of the lowest organisms, and so diverse in size and structure that they obviously belong to a number of species."

After these first epochal discoveries, Pasteur attacked the question of spontaneous generation, convinced that the theory did not fit the facts. To test his ideas, he devised an ingenious procedure, which has become known as the "swan neck" experiment. The prevailing view was, again, that germs would generate spontaneously in a meat- or vegetable-broth infusion or similar substance left open to the air. Pasteur poured a broth into flasks with an open mouth and a long curved neck that extended sideways from the flask, rather like a swan's neck or a backward letter "S" lying on its side. Some airborne particles would fall inside the open mouth of the flask, but gravity would keep the particles from moving up over the bend in the neck of the flask and then falling down into the broth. Pasteur showed that if the flasks were tilted so that some broth came into contact with the air at the end of the swan neck, and thereby became contaminated with airborne particles, germs would begin to grow in the broth. But if the flasks were not tilted, no germs grew. With this one experiment Pasteur exploded the myth of spontaneous generation, and proved that germs had to be carried from one growing place to another by some means.

Pasteur went on to a succession of scientific breakthroughs, including the process named in his honor, pasteurization, which keeps food from spoiling, and the first rabies vaccine. Together with major contributions from other scientists, such as Robert Koch in Germany, Pasteur's achievements laid the groundwork for the new field of microbiology and all that has flowed from it down to the present day, including the burgeoning field of genetic engineering. Remarkably, Pasteur, who died in 1895, did much of his greatest work after suffering a paralyzing stroke in 1873, when he was only forty-six years old.

Luckily for humanity, the significance of Pasteur's work was appreciated almost immediately. His studies of fermentation, for example, inspired the English surgeon Joseph Lister to campaign for antiseptic procedures in medical care and especially in hospitals. With Pasteur's data to back him up, Lister fared much better than Semmelweis. Queen Victoria made him a baron, and the mouthwash Listerine was named for him. Most important, operating rooms today are relatively sterile, germ-free environments thanks to his medical activism. (In fact, operating rooms are not entirely sterile; they contain low counts of up to about five germs per cubic foot of air.)

The new science of microbiology came along just in time. In Europe and America, the nineteenth century saw an ever-increasing amount of urbanization. As a consequence of rapid industrialization, the majority of people came to live in densely crowded city slums rather than on farms or in small towns and villages. In 1848–49 a cholera epidemic in London killed 14,600 people; another 10,675 died in the outbreak of 1854. The new, more crowded living patterns made it even more important to understand the nature of infectious germs and the ways in which they spread from person to person and group to group.

The knowledge gained by Pasteur and others helped to motivate massive public-works projects in the late nineteenth century. Paris, for example, was rebuilt from the underground up to provide clean water and sanitation. Similar efforts went forward in cities all around the world, but the biggest achievements were in Europe and North America.

Unfortunately, as I noted above, proper sanitation and reliably clean water have still not reached much of the developing world. Millions of people die every year because they don't have access to clean water or sanitary sewage systems. The World Health Organization estimates that it would cost $10 billion a year to bring clean water and sanitation to the 2.4 billion people who now live and die without them. That $10 billion is half of what U.S. consumers spend each year on pet food and about the same amount as European consumers spend on ice cream.

It bears repeating that in a time of global access from anywhere to anywhere, the world's public health has become everyone's concern. An epidemic that begins to rage in the shantytowns of an African, Asian, or South American city is only a plane ride away from Miami, New York, or London.

In these circumstances, we should support international efforts to improve sanitation and provide clean water for the world's poor. Here at home we should press for a national campaign to educate the public about germs and the need for good personal hygiene. Remember, eighty percent of infections are transmitted by touch, and germs are acquiring antibiotic resistance with blinding speed. As I'll show later, there is great hope for new antibiotics, vaccines, and other treatments down the road. But in the meantime, there is one vital thing we all can do now to protect ourselves and our families from unnecessary risks of infection: We can wash our hands properly, and much more frequently than most of us now do. This **Protective Response Strategy**, as I call it, is one of many I will share with you in the course of these chapters. For convenience, the index gathers together all the **Protective Response Strategies** at the end of the book.

PROTECTIVE RESPONSE STRATEGY: HOW TO WASH YOUR HANDS

Effective hand washing requires soap and water. Soap contains one or more surface-active agents, or surfactants. The molecules that make up surfactants have a water-attracting, or hydrophilic, end, and a water-repelling, or hydrophobic, end. As the soap dissolves in water, the surfactant molecules lower the surface tension of the water and make it better able to loosen particles on whatever is being cleaned. The surfactant molecules and water then hold the dirty particles in suspension until they are rinsed away.

The temperature of the water is not a critical factor in this process. But if the water is cold, it will not dissolve the soap as easily, and it may be so uncomfortable that you don't wash long enough.

To wash your hands effectively, wet your hands and lather with soap. Then rub the soapy water all over your hands and fingers, not forgetting to clean under your fingernails, for twenty or thirty seconds. Rinse, and repeat. This will not get your hands operating-room sterile, but it will remove any transient organisms you may have picked up since the last time you washed your hands.

For your health's sake, you should wash your hands several times during the course of the day. At a minimum you should do so before eating, after using a bathroom facility, and after contaminating your hands with a cough or sneeze. It's also advisable to wash your hands after shaking hands with someone. Although it's rarely possible to do so politely immediately after greeting someone with a handshake, you can avoid touching your face or mouth until you have the opportunity to wash. I also personally always wash my hands whenever I come into the house from outside.

In public bathrooms (I'll discuss proper toilet procedure in full a little later), you should wash your hands before using the toilet if you've had to touch a doorknob or other surface on your way in. When you wash your hands afterward, use a paper towel or a wad of tissue to shut off the faucet and turn the door handle. If there is no waste receptacle near the door, drop the paper towel on the floor. If enough people do so, there will soon be a receptacle there, as there should be.

3

The Germ Domain

In the Beginning Was the Germ

The farthest back that science can reach in the history of life on Earth is the germ. A fossilized bacterial cell in a rock found in western Australia dates to 3.5 billion years ago, and all other evidence agrees in pointing to microscopic single-celled germs as not only the smallest but also the oldest known living things.

Germs qualify as living things, along with plants and animals, because they share all the attributes that together form the scientific definition of life. Simply put, they are born and they die, and in between they engage in the three processes of metabolism, growth, and reproduction. Metabolism comprises all the physical and chemical processes involved in an organism's existence and activities, including the process of extracting energy from the environment to fuel growth and reproduction. Some germs act like plants in using photosynthesis to get energy from sunlight, converting carbon dioxide in the atmosphere into food and emitting oxygen as a by-product. Others act more like animals in feeding on some substance that they metabolize, or digest, with enzymes.

All living things on Earth also share one more feature. They are all made up of cells, from single-celled microorganisms to multicellular plants and animals. The most elementary units of life, cells are small bits of cytoplasm bounded by a thin membrane and containing nucleic acids (DNA and RNA), amino acids, proteins, fats, and carbohydrates. The average cell is about .0004 of an inch long. In the simplest terms, cells carry genes on their DNA, and the job of genes is to code precise sequences of amino acids, known as peptide chains or peptide sequences, into proteins, the building blocks of life. All living organisms are made up largely of proteins, and the formation of carbohydrates and fats is governed by proteins acting as enzymes. Every cell contains enough genetic information to reproduce itself, and single-celled creatures generally clone themselves by dividing in two. In addition, every cell of a multicellular life-form contains enough genetic information to make it theoretically possible to reproduce the entire life-form by cloning, a process that science has learned by imitating the behavior of single-celled germs.

Scientists are busy looking for evidence of life in outer space. They haven't yet found any, but living material basically consists of carbon, hydrogen, nitrogen, and oxygen. These fundamental elements of matter, which can form organic or inorganic compounds depending on what they are linked with and how they are linked, occur throughout the universe in stars and other objects. The essential precursors to life, organic compounds containing these elements existed before the planets in our solar system began to form. Using special radio telescopes, researchers recently identified one such compound, a sugar molecule called glycoaldehyde, in a gas and dust cloud near the center of the Milky Way. The planet Earth began life about 4.5 billion years ago as a hot ball of condensing gas and dust revolving around the sun, so it is not hard to imagine that glycoaldehyde or a similar sugar molecule, of stellar origin, could easily have formed ribose, the sugar backbone of the nucleic acids DNA and RNA. As Carl Sagan, the noted astronomer, put it, "All of life is made up of the same matter as the stars," and therefore the laws of the universe apply to all matter and all life, whether man or microbe. Or as Joni Mitchell wrote and sang, "We are stardust."

When the Earth was only a ball of gas and dust spinning in space, gravity and heat from radioactivity caused the layering of the planet. Heavy metals such as iron and nickel sank to the center, forming a molten liquid core. Medium-density silicates wrapped around the core to make the Earth's middle layer, known as the mantle. Everything else became the relatively active surface, the crust. At this time the Earth was so hot that water was present only in the form of vapor clouds, which completely surrounded the planet. Eventually the Earth cooled enough so that the water vapor condensed into liquid, which fell as rain. It rained so prodigiously for millions of years that the oceans were formed.

As yet the Earth's atmosphere had no gaseous oxygen, but comprised mostly methane, hydrogen, ammonia, and water vapor. How organic compounds were produced in this environment remains an open question. In the early 1950s, testing ideas first proposed in the 1920s by the Russian biochemist A. I. Oparin and the British biologist J. B. S. Haldane, two American researchers, Stanley Miller and Harold Urey, constructed a model of Earth's primitive environment in the laboratory. Miller and Urey showed that simply introducing an electrical spark into this model environment, as bolts of lightning were surely doing again and again in primitive Earth's real environment, could produce a variety of organic materials, including amino acids. Over the course of hundreds of millions of years, these organic materials would have accumulated in vast quantities in the Earth's primordial seas. Before the evolution of germs or other life-forms, there would have been nothing to break down or recycle them.

Research along these lines continues, as scientists try to establish more precise mechanisms for the formation of organic material. But recently another possible explanation has emerged. The early Earth was the scene of hellishly violent explosions as deep-seated rock and gases such as carbon monoxide interacted. It has been proposed that these interactions, catalyzed by abundant quantities of iron sulfide, produced a chemical called pyruvate, which is an essential component of all living cells. Pyruvate is the fuel for an energy-producing process, called the citric acid cycle, on which all living things depend, directly or indirectly. Pyruvates consist of three carbons and form the basis from which sugars and other carbon-based molecules are constructed.

Investigators have now successfully modeled these early rock and gas interactions in the laboratory, matching the success of the Miller/Urey experiments by showing how the first peptide chains might have been produced from a poisonous substance like carbon monoxide. There is support for the newer theory outside the laboratory as well. Some species of bacteria still thrive amid the rock and gas interactions that occur at volcanic vents on the deep ocean floor; they are the descendants of the ancient bacteria, or archaebacteria, to use their scientific designation.

Science cannot yet say which of these two theories holds more promise, whether both are correct (as seems likely), or whether an entirely different theory will be needed to explain the production of organic macromolecules such as peptide chains. But the next step in the evolution of life was the combination of these peptide chains into still larger molecules. This ongoing process eventually resulted in the emergence of the first prototypical cells, indirect evidence indicates, about 3.8 to 4 billion years ago. The scientific consensus is that the first of all living things, the first organisms capable of completing the full cycle of metabolism, growth, and reproduction, were prototypical blue-green algae, now known as cyanobacteria, and other single-celled bacteria.

The ancient blue-green algae, like their descendant blue-green algae today, lived by photosynthesis and thus produced oxygen. During this period of Earth's history, intense volcanic activity also produced oxygen, releasing it from the planet's interior. Over the course of perhaps three billion years, the oxygen produced by algae and the oxygen spewed out by volcanoes changed the atmosphere, creating an oxygen blanket that allowed higher life-forms to evolve from germs. But the existence of higher life-forms has never lessened the status of germs as the dominant creatures on the planet. From this point on, in fact, germs would never be less than a major influence in shaping the environment to support life.

For example, as oxygen produced by the photosynthesis of microorganisms rose into the upper atmosphere, it came into contact with ultraviolet (UV) rays from the sun. Exposure to UV light and to electrical discharges from lightning turns oxygen into ozone, which can then strongly absorb further UV rays. In this way, the Earth's ozone layer was formed. If land-dwelling organisms had to face the full force of the sun's UV rays, they would eventually be destroyed. The ozone layer in the upper atmosphere prevents this by acting as a protective shield. That is why scientists are currently concerned about the depletion of the ozone layer by pollution resulting from humanity's burning of fossil fuels and use of chlorofluorocarbons. In December 1998 the World

Meteorological Organization reported that the ozone hole that annually forms over the Southern Hemisphere had grown to ten million square miles, an area three times the size of Australia. This meant that the ozone layer was twenty-five percent more depleted than in previous years. By comparison, in 1981 the hole was only about 390,000 square miles in size.

The world is already seeing an increase in skin cancers and cataracts because the depletion of the ozone layer is exposing people to more UV light. The bottom line is that it has taken human beings only two or three hundred years of industrial activity and pollution to endanger the life-sustaining environment that germs produced over billions of years in Earth's early history and that they have sustained to the present day.

If There Were No Germs

To appreciate how essential germs are to life on Earth, let's try to imagine a world without them. Suppose that germs suddenly died out after higher life-forms had evolved from them. Before long, the remaining species would face three insurmountable challenges: no food, no oxygen, and no nitrates. Microscopic algae are the foundation of the entire world's food chain, which extends from these lowly germs to fish and other sea creatures and then on to land-dwelling organisms. If there were no germs in the seas, all marine life would starve to death, and land animals, including human beings, would eventually follow suit.

Similarly, the photosynthesis of land plants can provide only a fraction of the world's oxygen needs. Microscopic algae come to the rescue once again, supplying more than ninety percent of the world's oxygen with their photosynthesis. So if there were no germs in the seas, or if the ozone layer were to break down and let lethal UV rays hit the oceans directly, killing the world's marine algae in the process, the result would be that entire species, including human beings, would soon face extinction by suffocation. There would certainly not be time for evolution to adapt creatures to breathe a different sort of atmosphere.

And what about there being no nitrates, why would that be a

problem? The answer is that green plants must have nitrates in order to grow. And bacteria create nitrates by fixing nitrogen, as the process is called. Only a small amount of nitrogen exists naturally in the Earth's crust, but nitrogen gas abounds in the atmosphere. In the course of making nutrients for themselves, bacteria take this nitrogen gas out of the atmosphere and convert it into the nitrates that plants require.

The first bacteria to do this were the same prototypical blue-green algae, the earliest cyanobacteria, that gave the Earth its oxygen. Today's cyanobacteria continue to fix atmospheric nitrogen, insuring that plants can continue to flourish. You might say that these cyanobacteria, which stretch in an unbroken line of descent from the ancient archaebacteria, are four billion years young. They emerged as, and remain, the only truly self-sufficient creatures on Earth, in that they do not depend for their existence on any other living organisms. Instead, the rest of life depends on them.

Let's consider one more germless scenario. Suppose germs were not the basis of the world's food chain and its oxygen and nitrogen cycles, that their absence did not condemn other living things to extinction by starvation or suffocation. Suppose further that nothing like germs existed. As generation after generation of other plant and animal organisms lived and died, their bodies, their waste matter, and any other organic material they produced would begin to pile up on the surface of the Earth. Before thousands of years had passed, there wouldn't be an inch of ground left for plants to grow on or for other organisms to live on. The very oceans would be thick with corpses.

The second law of thermodynamics tells us that matter cannot be created or destroyed, only transformed. To keep the cycle of life going, nature needs organisms that will break down organic compounds into their fundamental inorganic elements, recycling them for future use. This above all is the role that germs fill in nature. For every naturally occurring organic compound, there is a species of germ which can break it down and make its inorganic constituent parts reusable. Without germs, all of the Earth would soon be one great garbage dump and graveyard, spinning lifelessly in space. You might say that without germs, humanity might have begun with Adam and Eve, but it probably wouldn't have lasted past Moses.

Germs' ability to break down organic compounds may well offer a way out of one of humanity's most pressing dilemmas, our competing needs for more and more energy and for ways to preserve the environment that our industries pollute with smog, oil spills, and other toxic compounds. For example, some species of bacteria feed on sludge and emit methane gas as a waste product. Methane can be burned as a fuel, just like the natural gas we use to heat our homes and cook our food. This means that bacteria could become the heart of combination sewage treatment/power plants, safely breaking down the waste of an entire city's population and supplying the city's energy needs at one and the same time.

Not only that, but methane-producing bacteria have counterparts called methanotrophs, which decompose methane and about 250 other toxic compounds. Greenways near streams contain methanotrophs that help purify the water. Imitating that natural process, government and industry have begun deploying other microbes to clean up oil spills, and similar efforts are sure to follow. If we are wise enough to learn the secrets that these germs can teach us, we can hope to bequeath a healthy planet to descendants of Adam and Eve who are yet unborn.

Animal, Mineral, or Vegetable?

Since the first ancient bacteria and cyanobacteria emerged 3.8 to 4 billion years ago, germs have branched out into a dizzying array of species. The known germ species number in the tens of thousands, and the total number of germ species currently in existence could number in the millions. Germs have adapted to profit from every quirk of nature. So-called hyperthermophiles can be found in the boiling hot springs of Yellowstone Park and near the lava vents of volcanoes, thriving at temperatures as high as 104°C (220°F). Cold-loving germs live in the snow of glaciers and undefrosted household freezers, and have been revived after lying dormant deep in the Antarctic ice for three billion years. Although many microbes live by photosynthesis and need light, others reside in total darkness in caves and on the ocean floor. Some germs find their niches in richly acid milieus, such as acid springs or mine waters, and some prefer highly basic ones, such as alkaline lakes. Aerobes must have oxygen, which will kill their opposite number, obligate anaerobes (they are obliged not to have oxygen). So-called facultative anaerobes can tolerate oxygen if necessary, however (they have the faculty of using or not using oxygen, according to circumstances).

Germs have not only colonized every part of the physical environment, land, sea, and air; they have also colonized every "higher" life-form as it has evolved. No new species ever emerges, in an environment shaped by and filled with germs, without its own complement of resident germs. As we'll see in the next chapter, human beings normally host substantial numbers of germs of many different species on their skin and in their respiratory and gastrointestinal tracts. We could not live without these germs, but we can become very sick when our relationship with them is thrown out of balance.

Science has faced an immense task in classifying a world's worth of germs. The catchall terms "germ" and "microbe" cover a vast range of microscopic algae, bacteria, fungi, protozoa, and viruses. The broad challenge in classifying germs is to decide whether particular species should be grouped with plants or animals, that is, as flora or fauna, or whether they should be categorized in some other way. For example, the cyanobacteria, or blue-green algae, have a clear link to plants in their common use of photosynthesis. But other germs have struck observers as being much more like animals. Indeed, protozoa means first ("proto") animals ("zoa"). After they are dead some germs even enter the mineral world, like the ancient protozoa whose shells form the blocks of stone that were used to build the pyramids.

The terms microbe and microorganism imply that all germs are too small to be seen with the naked eye. But scientists have recently discovered one exception to that rule in the seafloor sediments off the Skelton Coast of Namibia. There lives a bacterium that can grow to half a millimeter in diameter, about the size of a small grain of sand and over a hundred times the volume of any previously known large bacterium. The bacterium is so large because it has a storage sac that it fills with its food, nitrates, to tide it over during hard times. These large bacteria link up to one another in chains; because of their white storage sacs, they are informally referred to as "string of pearls" bacteria.

Since the nineteenth century scientists have devised several schemes for classifying germs. Some schemes divide all living things into only the two traditional categories of flora and fauna. Others employ as many as five categories to distinguish living things on the basis of cell type and other factors. These classification schemes aren't important for our purposes, but they explain one medical term that I'll be using frequently in what follows. Because the earliest attempts at classifying germs grouped them with plants, the germs that reside in and on human beings have come to be known as the natural human "flora." Medical science continues to use this term for convenience, despite the fact that among other germs human beings can host some thrity species of protozoa or "first animals."

The other special terms used to describe germs mainly come from their shapes as observed under a microscope. Many germs are rod-shaped, and both the Greek word "bacterium" and the Latin word "bacillus" mean "little rod." Other germs are spherical, and grow in chains or in bunches like grapes; the names of these germs, such as *Streptococcus* and *Staphylococcus,* often include the Latin word "coccus," which originally denoted a round kernel of grain. For their part, the words "spirilla" and "spirochete," based on the Latin word for spiral, are used to describe spiral-shaped germs. Finally, many germs are distinguished by their means of locomotion. Flagellate germs, for example, have whiplike flagellae at one end, which they swish back and forth to propel themselves. As I'll point out now and then in later chapters, the names of germs can contain intriguing information about their history or the particular ways in which they act.

Coexisting with Germs

The long evolution of life on Earth has produced a complex ecosystem, in which germs and other creatures live together in a variety of relationships. The interaction of two or more species in some ongoing way is called *symbiosis,* and there are three main categories of symbiotic relationship: *mutualism, commensalism,* and *parasitism.* Illness very often involves a shift in the relationship of two species from a mutualistic or commensalistic one to a parasitic one. The natural order is a system of checks and balances, a sphere of dynamic flux in which mutualism can give way to parasitism, and vice versa. In the case of organisms that cause infectious disease in people, the line between benefiting and harming their human hosts may often be very fine.

In a *mutualistic* relationship, each partner gives the other something that helps it to survive. A classic example of mutualism is the association of the common termite and the protozoa that lodges in its gut. The termite eats wood, but is unable to digest the wood on its own. The protozoa can digest wood, but only if it's been chewed up into small pieces. Neither species could survive alone, but together they make a good living wherever there is enough wood.

In a *commensalistic* relationship, one organism gains something from another without giving something back, but also without doing any harm. A textbook example of the "taker" in this kind of relationship is the clownfish, which lives protected from predators within the waving mass of a sea anemone's poisonous tentacles. On closer examination, however, it seems that the clownfish is not entirely a freeloader. It attracts other fish on which the sea anemone can feed. Very likely, many of the relationships we have labeled commensalistic are really mutualistic ones, in which we haven't yet recognized all the ways that each party benefits.

In a *parasitic* relationship, one organism derives benefit from another and in so doing harms its host. For example, the tapeworms that come from beef *(Taenia saginata)* and pork *(Taenia solium)* can grow many feet long in our digestive tracts, gorging themselves on what we eat. They not only rob us of nutrition, but also cause severe anemia, intestinal obstruction, and sometimes even death. Likewise any infectious disease could be described as a germ acting as a parasite on a human host.

One of the most fascinating aspects of parasitism is that lethal parasites can gradually moderate their impact on their host species, at least up to a point. After all, it makes no sense for a parasite to kill its host. If the host dies, the parasite loses food and shelter and has to move on in search of its next meal and a new roof over its head. Thus the parasite that causes toxoplasmosis, a potentially fatal disease that can produce lesions in the eye and the nerve cells of the brain, has adapted so well to living in human beings that it usually doesn't kill or even hurt them. All other things being equal, a person with untreated toxoplasmosis can live out a full life span, and this is to the parasite's advantage. The condition only becomes a serious problem in people with suppressed or damaged immune systems. Pregnant women are by definition immunosuppressed, and so they are routinely screened for toxoplasmosis. If a woman does not have antibodies to toxoplasmosis from previous exposure, and if she is then exposed to it during the course of her pregnancy, the fetus can be endangered. Prenatal toxoplasmosis is the major cause of congenital blindness and other birth defects. Congenital infection can even lead to stillbirths.

Toxoplasmosis is the number-one parasite worldwide. Many people pick it up from their pets, especially cats that get it from eating mice and birds. Indoor cats are much less likely to have toxoplasmosis, but any cat may have it. Cats pass the oocysts of the toxoplasmosis parasite in their feces, and people can inhale the oocysts just by getting too close to the feces. For this reason, pregnant women are warned never to clean a cat's litter box. The cysts may also be present in the raw or undercooked meat of other animals. At least two billion people around the world harbor the toxoplasmosis parasite; most don't know they have it and will never by troubled by it. Almost one hundred percent of the French population, for example, has contracted toxoplasmosis from eating steak tartare and other raw meats.

The same thing is true of other parasites, such as HIV, the virus that causes AIDS. Although this virus, in one form or another, has existed for a long time in monkeys and other primates, which can live full life spans with it in their blood supplies, it is very new to human beings, who cannot survive it except in rare cases of individuals with an as yet unexplained anomaly in their immune systems. But if HIV gets enough experience with human hosts over a sufficient number of generations, it will almost certainly mutate into less deadly, but still very harmful strains, so that it can enjoy a longer life in each person it infects. Parasites are sort of like Freddie the Freeloader. They don't want all your money at once; instead they'd prefer to borrow from you forever.

As I've already suggested, there is often a fine line between mutualism and parasitism. For example, the bacterium *Streptococcus pneumoniae* can give people pneumonia if it gets a foothold in their lungs. Yet the same germ resides benignly, elsewhere in the respiratory tract, in fifteen percent of the population. In these cases the bacterium has become part of the individuals' normal flora, and lives within them in a mutualistic state. But if something conveys that *S. pneumoniae* into the same individuals' lungs, it could give them pneumonia too, and the bacterium would then be living inside them in a parasitic state. Another good example is *Staphylococcus aureus,* a mutualistic component of most people's normal skin flora. But if *S. aureus* gets into an open wound site, as can happen to patients after surgery, an infectious illness could occur and the germ would then become parasitic.

In the next two chapters I will show you how germs develop and maintain their relationships with us. When these relationships are in balance, germs protect us from disease and perform critical tasks such as helping us to digest our food. When they are out of balance, we can all too easily take sick and even die. The more we understand about germs, however, the more we'll be able to coexist with them in a healthy way.

4

The Germ Factory

For microbes, the human body is like a planet unto itself, with a range of habitats almost as diverse as those of the larger world. Although some germs can exist under arid conditions, most thrive on humidity, which is abundantly present in the moist channels of the human body. Over the course of human evolution, many germ species have become mutualistic homesteaders in the human body, earning their food and shelter by performing life-sustaining tasks for their hosts. For example, like the termite with wood-digesting protozoa in its gut, human beings also house germs, such as *Escherichia coli* (a.k.a., *E. coli*), that help break down food and assist in the formation of essential vitamins and nutrients.

Some germs do their human hosts a vital favor by actively repelling dangerously parasitic germs, or simply by taking up space that the dangerous germs would otherwise invade. In the nose and throat, *Streptococcus viridans* wards off potentially pneumonia-causing *Streptococcus pneumoniae*. The bacterial species *Neisseria* can stimulate the body to produce an antibody shield against invading meningitis germs. And our skin is a veritable picnic blanket for microbes, notably the organisms *Staphylococcus aureus, Staphylococcus epidermidis,* diphtheroid bacilli, and *Propionibacterium* sp, which not only patrol for enemy strains but also keep the skin from being smothered in debris by feeding on oils and dead cells. These microbes and many others constitute human beings' normal flora.

Every germ has optimal physical-chemical requirements and seeks the ideal environment in which to grow and reproduce. Some of the ecological conditions that affect a germ's comfort level at a particular site in the human body, or elsewhere for that matter, include pH (the acid/base proportions), moisture, temperature, and other growth factors such as the level of available nutrients; the presence of supportive or detrimental germs nearby; and even the influences of magnetic fields. No single factor will determine the site's suitability, any more than a house's size alone would determine its attractiveness to a potential buyer. But a single factor, like the proximity of a house to a noisy freeway, can make a site inhospitable to prospective germ residents. Mindlessly responding to all the ecological conditions involved, germs ceaselessly seek out and create supportive niches for themselves. In the process, they gain competitive advantages over other home-seeking microbes.

Consider the ecological conditions of the skin. Human skin presents an acidic environment, with a pH of about 5.5. It is protein rich, because it is covered in sloughed-off, or desquamated, cells. It exudes oils, with some people's skin being more oily than others'. Finally, there is plenty of oxygen available on the surface of the skin, but none beneath the surface in its follicles. So the microorganisms that will successfully colonize the skin will be proteolytic (protein eating), lipophilic (utilizing lipids or oils), acidophilic (preferring an acid environment), and versatile enough to live either aerobically or anaerobically (with or without oxygen). The germ species that best fit the bill are *S. aureus, S. epidermidis,* diphtheroid bacilli, and *Propionibacterium* sp, and indeed these are the skin's main indigenous inhabitants.

To establish themselves in favorable sites, germs rely on specific attributes of their structure and functional abilities. One of the most important such attributes is "adherence affinity," the ability of a germ to attach to specific kinds of cells and tissues. For example, certain germs have high adherence affinity with—that is, they stick well to—the cells lining the mucous membranes of the upper respiratory tract. Because they attach so well to these membranes, the germs gain a competitive advantage over other microbes that might be attracted to the same site because of its moisture, among other appealing factors. The respiratory microbes might in turn find themselves attracted to damp places elsewhere in the body, such as the bladder, but be unable to lodge there because they do not have sufficient adherence affinity for these surfaces. This precisely describes the situation of group A beta streptococcus (a.k.a. *Streptococcus pyogenes*), the common agent of strep sore throat, among other ailments. It adheres well to the mucous membranes in the respiratory tract, but not to the tissue surfaces in the bladder.

On the other hand, *E. coli* adheres very well to gut tissue and thus has established itself as one of the chief indigenous intestinal germs. But because *E. coli* lacks a special adherence affinity for cells in the mouth, it cannot compete effectively for a position there, no matter how attractive the site might be for other reasons. The germs that can attach themselves firmly to specific body sites tend to dominate those sites and prove difficult to dislodge or eradicate.

The competition among germs for dominance at specific body sites exists in dynamic flux. Anything you do to your body can affect the relative status of your resident germs. Using deodorants, douches, skin creams, and perfumes changes the physical-chemical conditions wherever they are applied. This may give different sets of germs the upper hand and lead to rashes or other unwanted conditions. For example, using too much antibacterial deodorant can kill off normal skin germs and allow for overgrowth of yeast, giving rise to a yeast infection. Likewise, anything you ingest can also affect your body's natural flora. Antibiotics, for example, can often disturb germs at sites that are not the medicine's intended target. Most people have experienced side effects like upset stomachs and diarrhea that occur when antibiotics, taken for infections elsewhere in the body, have disrupted their normal gut flora. Luckily, these side effects are usually temporary, and the body's resident germs regain their normal balance after the antibiotic therapy ends.

Human beings are also in a state of dynamic flux with regard to daily challenge by a wide variety of germs in their environment, including especially their contacts with other people. A seesaw relationship exists between human beings and germs, whether the germs naturally reside within them or outside them. Most people with healthy immune systems can readily ward off most challenges from microbes. And although some members of the population—children, pregnant women, the elderly, and the immuno-suppressed—will always be at greater risk of infection, it is important to remember that the overwhelming majority of the germ-related illnesses that do occur can be prevented with proper hand washing and other simple precautions. Later in this chapter, I will explain how both benign and dangerous germs find their way into human hosts and how they move from one body site to another, from person to person, and from population group to population group. In all three cases, the process can easily turn a benign, mutualistic germ into a dangerous, parasitic one. In fact, the most common infections in people result from the transmission of normal human flora of the skin, feces, and respiratory secretions to a place in the body, or in the body of another person, that cannot host them in a mutualistic way.

Location, Location, Location

Although other factors are critically involved, the key to understanding how human beings can coexist with their resident germs, like the key to buying good real estate, is location, location, location. Germs in and on the human body reside in two sorts of areas that can be represented schematically by two concentric circles, A and B.

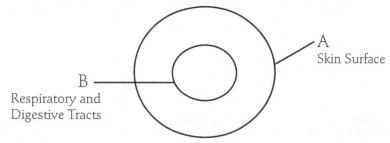

The outer circle, A, represents the surface of the skin and all of its invaginations such as the nostrils, conjunctiva (the mucous membrane on the inside of the eyelid and the outer surface of the eye), vaginal vault, and ears. The inside area of the inner circle, B, represents the respiratory and digestive tracts: the mouth and sinuses, throat, stomach, intestines, and colon. The area between the circles represents the internal organs—heart, lungs, kidneys, liver, etc.—which are normally germ free. If the sterility of this germ-free area is broached, very serious illness can result. Although the lungs may seem particularly susceptible to germs inhaled from the air, whatever microbes you breathe in are usually engulfed and destroyed by the body's immune defenses, or simply coughed up, in short order.

The largest concentrations of germs reside in two places: the colon, where the feces are formed, and the gums. Both of these areas contain about 10^{12}, that is, about one trillion, bacteria per gram. These bacteria are predominantly anaerobic. Growth on tooth surfaces follows, with concentrations as high as 10^{11} or 10^{12} bacteria per gram of tartar and roughly equal numbers of anaerobic and aerobic bacteria. Saliva, also with an equal ratio of anaerobes and aerobes, contains about 10^8 bacteria per milliliter. The large quantities of germs in the mouth and feces make bites and fecally contaminated wounds especially vulnerable to serious infections.

The nose contains many fewer bacteria, about 10^3 to 10^4 bacteria per milliliter with a 2:1 ratio of anaerobes to aerobes. The skin, for its part, has about 10^5 (100,000) bacteria per gram, with most skin areas having a 100:1 ratio of anaerobes to aerobes. These bacteria reside either on the surface of the skin or within its two million or so pores. As I mentioned in Chapter 2, skin is the largest single mechanism for transferring germs from person to person, with approximately eighty percent of all infections being transmitted by contact.

The following table outlines the number of germs and the ratio of anaerobes to aerobes at different body sites.

NORMAL FLORA OF HUMANS

Body Site	Number Per Milliliter Or Gram	Ratio Anaerobes/Aerobes
RESPIRATORY		
Nose	10^3–10^4	2:1
Saliva	10^8	1:1
Gingival crevice	10^{12}	1000:1
Tooth surface	10^{11}	1:1
GASTROINTESTINAL TRACT		
Stomach	10^0–10^4	1:1
Small intestines	10^4–10^7	1:1
Colon (feces)	10^{12}	1000:1
SKIN		
Surface	10^5	100:1
Vagina	10^9	4:1

As the table illustrates, the human body is a veritable germ factory. Each body site not only contains an immense quantity of germs, but a diversity of species as well. There are over five hundred different varieties of germs in human feces, more than two hundred in the oral cavity, about thirty in the vaginal vault, and lesser amounts in other body areas. Most of these germs are different species of bacteria, but the normal human flora can also include some fungi (molds and yeasts) and protozoa (one-celled animals).

How Germs Spread

Human beings emerge from the womb sterile, germ free. But as a baby passes through the birth canal, it picks up a germ passport, a complement of microbes that will seed the growth of its normal flora. From that point on in a person's life cycle, the normal flora develop in an orderly sequence called ecological succession, which eventually results in a generally stable population of diverse germs at various body sites. As a child grows and matures, its body chemistry changes, making it hospitable to a fuller range of normal bacteria. This process is part and parcel of the development of a healthy immune system; most childhood diseases occur before the age of six precisely because the first six years of life are the time when most of the normal flora are establishing themselves in a child's body. A second important stage occurs at puberty, when hormonal changes stamp one's germ passport with a new visa, as it were, opening the door to new microbes that enhance the functioning of the adult body. This set of germs will in turn phase out to some degree in old age.

This progression can be seen especially clearly in the normal flora of the mouth. Germ-free at birth, an infant's mouth takes only three to five days to nurture its first population of *Streptococcus salivarius,* a germ picked up from the mother's birth canal or skin, which then lodges in the mucous membranes of the cheeks and tongue. Six to nine months later, as teeth develop, *S. mutans* and *S. sanguis* crop up to colonize them and join *salivarius* as resident flora; their presence protects against infection by less benign germs.

Infants come into contact with *S. mutans* and *S. sanguis* almost immediately, from the skin and respiratory secretions of the people who touch them, talk to them, and coo over them. But the germs have no niche to fill, and thus do not take up residence and propagate in any numbers, until the teeth come in. In old age, as the teeth fall out, *S. mutans* and *S. sanguis* begin to fade away, but they will spring back in full force if their host starts using dentures.

Scientists analyze the movement of germs, whether benevolent or harmful, in terms of what is called the chain of infection. Studying an infectious agent's progress from person to person and population group to population group is called epidemiology. Tracking an epidemic literally means tracking what is on people (from the Greek words "epi," on, and "demos," people). The first link in the chain of infection is the germ itself, called the *agent*. The agent may be a bacterium, virus, fungus, or protozoa that is found at some source, known as a *reservoir*. The reservoir could actually be a body of water, but it could just as easily be an animal or plant that is already home to the germ, or even an inanimate object that an infected person has recently touched. Some route of *transmission* then transfers the agent to a new *host* organism, completing the chain and giving the germ an opportunity to propagate in a new place. Infectious disease may or may not result at this point; this depends, as we'll see in the next chapter, on the virulence of the germ and the health of the host, among other circumstances. There are four possible routes of transmission: contact, common vehicle, airborne, and vector-borne.

Contact Spread

For *contact spread* the prospective host must have actual contact with the source of germs. Contact can be direct, indirect, or by droplets. *Direct contact* involves a person-to-person spread. There must be some degree of actual physical contact between the source of the germs and the new host. Shaking hands or kissing the face of someone with a cold can easily transfer the cold virus to you. Sexually transmitted diseases such as syphilis, gonorrhea, HIV, and chlamydia are other good examples of direct-contact spread.

Indirect-contact spread is distinguished from direct-contact transmission by the passive involvement of an intermediate object, usually inanimate. Rhinoviruses (cold viruses) and caliciviruses (vomit- and/or diarrhea-inducing viruses that are also known as the stomach flu) can make you sick, for example, if you touch a doorknob that a contagious person has touched very recently.

Or, if not a doorknob, try a football. Recently a large number of players and personnel on the Duke and Florida State football teams fell sick with vomiting, nausea, stomach cramps, and diarrhea. Because so many people became ill, an investigation was conducted. The only interaction between the two teams was during and immediately after the game, but as many an announcer has said in describing a vicious tackle or other collision, football is a contact sport. Players are constantly touching each other and the ball. As they catch their breath between plays, football players often take their protective mouthpieces out, transferring any germs present in their saliva to their hands. This would be more than enough to account for the sharing of germs between the two teams during the course of play or in shaking hands afterward. But video footage of the game also showed some Duke players vomiting on the sidelines, getting germ-rich vomit on their jerseys, and then grappling with Florida State players.

Sure enough, the illness had begun on the Duke side of the field. The day before the game, the Duke players had all eaten a box lunch of turkey sandwiches prepared by a food worker who was sick with a Norwalk virus, one of the caliciviruses I just mentioned that are known as the stomach flu or the "twenty-four-hour virus" because they cause nonbacterial gastroenteritis that usually lasts for only a day or so. Norwalk viruses are very hardy; one study found that a Norwalk virus can survive on uncleaned carpet for a month or more. When researchers took cultures from the sick members of the two football teams, they found the exact same strain of Norwalk virus as they found in a culture taken from the person who made the turkey sandwiches, proving that the cases all formed one chain of infection.

Now let's tally the final score: One food handler infected forty-three Duke players, because they ate the sandwiches, and these players in turn infected eleven Duke coaches and other personnel and eleven Florida State players, because they touched each other. So with all the big, strong, well-conditioned athletes on the field, that makes a grand total of sixty-five tackles for the game's hardest-hitting participant, a microscopic Norwalk virus. A first in sports history! (At least, the first to be documented so rigorously.)

The final form of contact spread is via droplets. *Droplet spread* refers to the brief passage of the infectious agent through the air when the source and the victim are relatively near each other, usually within several feet. This means of transmission can involve coughing, sneezing, and even talking. Droplet-spread infections include *Streptococcus pneumoniae* (a common cause of pneumonia) and viral respiratory diseases such as influenza.

One of the most important mechanisms of contact spread is known as the fecal-to-oral route. This occurs when an infected person has failed to practice adequate personal hygiene after defecating. The contact can be direct, as by shaking hands, or indirect, as when you eat food that an unhygienic germ carrier has handled. The fecal-to-oral route is to blame for much of the food poisoning that people get from salad bars, sandwich counters, and restaurants (or turkey sandwiches in box lunches!). Hepatitis A, among other nasty germs, can be passed from one person to another by the fecal-to-oral route.

This underlies the need for an aggressive public-hygiene education campaign, such as I've already talked about in terms of hand washing. This campaign should include the proper way to wipe one's anus after defecating, another **Protective Response Strategy.** Different degrees of wiping are necessary, according to the consistency of the stool. The better formed the stool is, the less fecal debris there will be on the anus. To wipe yourself hygienically, first use soft dry toilet tissue to wipe away as much fecal contamination from the anus as possible. Follow this step with a series of wet wipes using a saline or a soap-and-water solution. There are numerous commercial products available for use in wet wiping. You can make an effective saline solution at home simply by putting a spoonful of salt in a glass of water. Plain water should not be used for this purpose, unless nothing else is available, because it can irritate the mucosal tissue of the anus in some people. Continue wiping until there is no obvious soilage on the toilet tissue. You may alternate between wet and dry wiping for greater effectiveness. Then pat dry, and be sure to wash your hands as outlined at the end of Chapter 2. Anal pruritis, or itching, may indicate improper wiping technique, among other things.

A fascinating aspect of the fecal-to-oral route deserves mention. If feces contains no infectious-illness-causing germs, and if you have a healthy immune system, you can literally spoon-eat it without a problem (except, of course, aesthetically!). In the first place, a person eating his or her own feces would simply be reintroducing germs that were already present in the digestive tract in huge numbers. And even though fecal matter contains the body's highest concentration of germs—again, this is about one trillion bacteria per gram—the microbes from someone else's feces would have a tough time competing with a healthy person's normal gut flora. Some animals, such as rabbits, must eat their own feces, because the feces contains nutrients that they cannot absorb in any other way. Fortunately, this is not the case with human beings.

A related concern is the ingestion of germs during sexual activity, from kissing to fellatio and cunnilingus. It is easy to see how some fecal germs might contaminate the nearby area of the genitals, and in women the vaginal vault is host to a number of germ species, as we'll see. But assuming that neither partner has a sexually transmitted disease or other overt infection, there is no danger of infectious illness from ingesting germs during sexual contact. Any ingested genital or fecal organisms will enter the gastrointestinal tract, where they will either be destroyed by the acid environment in the stomach or pass into the large intestines to join and compete with the very large microbial population of normal flora there.

Common-Vehicle Spread

Common-vehicle spread involves a contaminated inanimate vehicle that serves as the vector for transmission of the agent to several or many persons. Common-vehicle spread can become a major news story when packaged food is contaminated with *Salmonella* or *Campylobacter* or a public water supply is contaminated with parasites, all of which can cause vicious bouts of diarrhea and nausea, with or without vomiting. Members of at-risk population groups, such as children, the elderly, and the immunosuppressed, may even die as a result of being infected in this way.

Airborne Spread

Airborne spread implies the airborne spread of germs over a distance of more than several feet between the source and the victim or victims. The infectious organisms are usually contained within droplet nuclei which are five microns (five millionths of a meter) or smaller in size. These particles can remain suspended in the air for hours or days, depending upon environmental factors such as humidity and wind currents. If you've stood downwind from a cigarette smoker and inhaled the smoke, you have experienced airborne spread. The classic example of airborne spread is the transmission of the tuberculosis bacillus by means of droplet nuclei. The deadly anthrax bacillus can also spread in this way; later, I will discuss how a terrorist might try to take advantage of airborne spread in order to infect a population with anthrax. Organisms may also travel in dust. In one recent case, a family kept coming down with salmonellosis, which can cause debilitating vomiting and diarrhea, because of contaminated dust in a vacuum-cleaner bag. Most vacuum cleaners actually recirculate a considerable portion of the dust they suck up; so the members of this family were unknowingly resuspending the infectious *Salmonella* organisms in their household air every time they turned on the vacuum cleaner.

A commercial airliner can act like a leaky vacuum cleaner. Airlines are supposed to keep the air in their planes clean by recirculating it through high-efficiency particle filters, which remove most viral and bacterial germs, and mixing it with fresh outside air. But airlines have stinted on these measures in the past, and may still do so, despite government regulations that require them. The longer the flight, naturally, the greater the danger that passengers will be infected by the respiratory secretions of a sick fellow passenger being constantly recirculated through the cabin. In 1994 the Centers for Disease Control, in Atlanta, reported the spread of tuberculosis on a flight from Chicago to Honolulu. Six passengers contracted TB while sitting helplessly strapped in their seats; four of them had no other risk factors for the disease, except that they were only two rows away from the sick passenger who carried the TB onto the plane. Although the chance of getting TB in this way is small, the World Health Organization says it is still a concern on flights of eight hours or more.

Vector-borne Spread

There are two types of *vector-borne spread*. One involves the mechanical transfer of microorganisms to a host from the body or appendages of another organism. Flies with *Shigellae* or *Salmonella* on the outside of their bodies can transmit these germs simply by alighting on a host organism (or food stuff). The second kind of vector-borne spread occurs when an infected animal bites another organism. Mosquitoes are the vector for transmitting the malaria parasite, which they carry inside their bodies, to human hosts. In the same way, ticks can transmit rickettsial diseases such as Rocky Mountain spotted fever, and Lyme disease.

It is important to remember that the final link in any chain of infection will be the host. As I'll explain in more detail in the next chapter, healthy hosts can be "infected" with even quite nasty germs, in the limited sense of picking them up by contact, common-vehicle, airborne, or vector-borne spread, and yet not get sick because their immune systems or other factors prevent the disease-causing germs from gaining a firm foothold.

The environment can significantly influence the chain of infection. Some environmental factors can influence all the links in the chain of infection, whereas others have more limited impact. For example, high humidity favors the growth of most germs, whereas desiccation, or drying, hinders it. Yet extremely dry air dries out a host's mucous membranes and makes them less able to protect against infection. Humidity thus can affect the viability of a germ agent at its source, its transmission through the air, and the effectiveness of a host's resistance and immune response.

Some ecological circumstances affect germs in a very specific way. The growth of a germ in its reservoir, or source, may depend on particular substances in the soil or water. A germ's survival at its reservoir, or during transmission, may also depend on temperature, pH, substrate, ionic composition, or even radioactivity. Toxic substances in the environment, perhaps ones produced by other germs, may keep a germ down. *Streptococcus pneumoniae* bacteria emit a substance that inhibits the growth of *Staphylococcus aureus* when both exist at the same body site. This adversely affects transmission of *S. aureus,* but favors that of *S. pneumoniae.*

Airborne transmission can be influenced by the speed and direction of air movement. A tuberculosis patient is more likely to spread the disease to other people when he or she is confined to a room rather than being outdoors. The enclosed space will allow dangerous concentrations of the tuberculosis bacilli to build up as the patient exhales. Common colds are also more frequent during the winter months because people tend to stay indoors with the windows closed, which reduces air circulation. In the summer, air-conditioning will circulate some fresh air into a room even if the windows are closed.

The temperature can play a major role in outbreaks of disease associated with common-vehicle transmission. *Staphylococcus aureus* growing in a Boston cream pie or potato salad will produce larger quantities of toxin during a hot summer picnic than during a fall tailgate party at a football game. Environmental conditions can also affect the movement of vector-borne organisms. If one could shrink the breeding grounds for mosquitoes through better drainage in urban areas, for example, one could lessen the incidence of diseases like malaria and West Nile fever, and antimalaria campaigns often feature such efforts. Unfortunately, controlling the mosquito population has proven to be a very difficult task, as we shall see.

Another element in the movement of germs is the individual means of locomotion that different germs possess and that they use to move on and within their hosts, as well as from host to host. Many germs are flagellate—that is, they use a whiplike appendage called a flagellum to propel themselves within a medium such as water or the bloodstream. Other germs use cilia, tiny hairs, to grab on to things and pull themselves along a surface. Still others employ pseudopodia, "false feet," which means that they protrude out at one edge and then pull the rest of themselves along, rather like an inchworm. Some germs go through a stage where they take the form of spores or cysts that can be lofted along in the air. It is not uncommon for the same germ to have successive life stages in which it becomes in turn a spore, a flagellate, or a pseudopod to take advantage of prevailing conditions in its environment or host organism.

As we'll explore in the next chapter, the complex environmental conditions within the human germ factory may become vulnerable to sudden shifts. In the case of polymicrobial infections, involving two or more disease-causing germs, one infection may clear the way for another more serious one. Suppose you are among the fifteen percent of the population who carry *St. pneumoniae* in your normal respiratory flora. A rhinovirus, which causes the common cold, might give that *St. pneumoniae* a chance to get from the nose into the sinuses, where it can give you sinusitis. Or the *St. pneumoniae* might piggyback along with some rhinovirus secretion that you inhaled into your lungs, making you vulnerable to a bronchial infection or even pneumonia. If this were to happen to you in older age, when your immune system had become less vigorous, antibiotic treatment might not be able to save your life. The rhinovirus lights the match, so to speak, but the *St. pneumoniae* burns the forest down.

Yet another possibility is that a germ may attack different parts of the body in succession in quite different ways. For example, a person sick with a stomach virus might, in the course of vomiting, convey virus particles into the eustachian tubes and then into the labyrinth of the inner ear. The inflammation that can result, viral labyrinthitis, is hardly fatal, but the dizziness and vertigo it induces by disturbing the body's balancing mechanisms can be debilitating long after the gastrointestinal upset has run its course.

The potential for polymicrobial and multiple-site effects means we have to be very alert for unusual or especially long-lasting symptoms, even if we get sick from an everyday cold or flu. Whenever your symptoms seem out of the ordinary, consult your doctor. That labyrinthitis might be viral, in which case you just have to wait until it subsides, or it might in rare cases be bacterial in origin, in which case you will need antibiotic treatment as quickly as possible to protect against permanent loss of hearing and balance.

To sum up, the chain of infection will be different for different varieties of germs and infectious diseases. To lessen the toll that infectious diseases inflict on our communities, we must carefully study each link of the chain. Only that way can we devise appropriate methods of control and prevention to add to the always essential strategies of good public and private hygiene.

The (Not So) Sweet Smell of Human Flora

The germs that make up the human flora are, of course, too small to be seen. And except for producing intestinal rumbles and gurgles now and then, as they break up food and release gases, they cannot be heard. But all too often, they can be smelled. We're all familiar with the unpleasant smells that anaerobic bacteria produce as they grow in sewage, rotten food, the decaying corpses of animals, and stagnant bodies of water. The anaerobes in the human body also emit foul-smelling by-products, and the heavy concentration of these germs in the mouth and fecal flora accounts for the odors associated with bad breath and feces.

Bad Breath: Its Causes and Cures

Aside from an infection such as periodontitis or some other physiological problem, chronic bad breath has three potential microbial sources: odors emanating from the back of the tongue, from degraded food particles trapped between the teeth, and from the stomach.

Chronic bad breath is different from so-called morning breath. Morning breath, more accurately known as temporary bad breath, is a normal consequence of the waxing and waning of anaerobic mouth flora. When you wake up in the morning, the concentration of bacteria in your mouth is at its highest point. The action of brushing your teeth or chewing your breakfast reduces the concentration. By lunch time, the relative inactivity in your mouth has let the concentration rise again. Eating lunch reduces the concentration once more, and the cycle continues through dinner and the nighttime hours. Brushing your teeth, gargling, eating, chewing gum, and sucking on candy will all reduce the heavy colonization of germs in the mouth and serve as an effective **Protective Response Strategy** for temporary bad breath.

In contrast, chronic bad breath typically results when specific circumstances encourage a persistent heavy concentration of anaerobic bacteria in the mouth. Some people are plagued with a relatively high population of germs on the back of the tongue, where they easily survive ordinary dental hygiene. These germs produce malodorous sulfites. In order to eliminate odor caused by such tongue bacteria, a person can gargle with a hydrogen peroxide solution (one part peroxide to one part water). The peroxide oxidizes the sulfites into odorless sulfates, eliminating bad breath for as long as twelve hours. Gargling once in the morning and once before bedtime should do the trick for anyone who suffers from this problem.

A second common source of bad breath is food particles, especially meat fibers, caught between the teeth. Anaerobic digestion, that is, putrefaction, of meat protein can cause bad odors to emanate from these trapped particles. The solution here is easy: Use a water pick or dental floss daily to clean between the teeth. A somewhat less efficient alternative is the good old-fashioned toothpick.

Finally, various digestive problems, eating heavily spiced foods, or a combination of both, can waft odor vapors up from the stomach to exit via the mouth. An old-fashioned but very effective remedy, one I learned from my grandfather, is to chew several wads of parsley after lunch or dinner, especially after eating foods rich in spices like garlic. Over-the-counter digestive aids, commonly available in gel tablet form, achieve the same effect with oils from parsley and celery seed. If you have chronic bad breath and one of these three remedies does not help, you should see your dentist or physician. You may have a more serious problem than you think.

One little-known area of high bacterial growth called a food lithe, containing about a trillion bacteria per gram, can also produce a foul odor. About twenty-five percent of people who do not have their tonsils surgically removed have a harmless anatomical anomaly that allows bacteria and food to accumulate more easily within the tonsils' fissures. The bacteria build an outer protein capsule that encases the bacteria and food particles in pea-sized structures. About a dozen different anaerobic bacterial species live inside the structures in a complex relationship that scientists have not yet been able to duplicate in the laboratory. When the pea-sized structures reach maturity, they are swallowed with food or they get coughed up or spit out. If one were to squeeze a food lithe, it would emit a very unpleasant anaerobic smell.

Foot Odor

Fungi and bacteria can both cause foot odor. Most people are familiar with the smelly feet of an "athlete's foot" victim. This infection of the skin can result from contact with any of several different species of fungi called dermatophytes. Three dermatophytes commonly found in gym and sports club showers and locker rooms, among other places, are *Epidermatophyton, Trichophyton,* and *Microsporum.* Less frequently, another type of fungus, *Candida albicans,* a yeast, can cause similar infections. As a **Protective Response Strategy**, be sure to wear clogs or similar footwear when showering in a gym or health club.

These fungi can all grow on the skin, hair, or nails, resulting in skin peeling (desquamation) and eventual decomposition of tissue, with an accompanying bad smell. The infections are usually superficial and do not invade deeper tissue. But the acute, swollen lesions they can produce are painful, as any athlete's-foot sufferer can attest. For reasons that are still unclear, men are infected at a greater rate than women. The disease is also more common in hot weather, or whenever excessive sweating or wetting of the feet occurs. Every human foot has more than 125,000 sweat glands, but some people sweat much more copiously than others. In general, the greater the amount of sweat produced, the greater the risk of a foot odor problem.

Sweating plays a critical role in bacterial foot odor, which is actually more common than fungal athlete's foot. In this case, the odor results from the normal skin flora bacterium *Staphylococcus aureus* growing in unusually high numbers on the feet of individuals who sweat prodigiously and who tend to wear black- or blue-colored synthetic socks. The physical-chemical conditions created by the sweat and the dark synthetic fibers create an ideal environment for *S. aureus.* Strictly speaking, this process is not an infection at all. Instead, the dark dyes in the sock fibers encourage a population explosion in a colony of normal human flora.

If you have this common condition, you can treat it very easily. Simply wear light-colored all-cotton socks for a while, and be sure to dry your feet carefully before you put on the socks.

Why Some Babies Smell Different from Others

I'm sure you've heard the expression "You are what you eat." This is emphatically true of babies. The feces of babies fed with cow's milk and formulas smells much stronger than that of breast-fed babies. Cow's milk has more than twice as much protein as breast milk and is also richer in calcium. The increased quantity of these substances causes babies to produce more feces than breast milk does and to grow bacteria in their intestinal tracts that are more like those of adults. The combined result is a stronger smell. Likewise, formula-fed babies also grow germs in their intestinal tracts—*Bacteroides, Clostridium,* and others—that are similar to those found in an adult's normal flora. In contrast, several studies suggest that breast-fed babies are more often colonized by beneficial anaerobic bacteria called *Bifidobacteria,* which impart a more pleasant odor to feces. Eventually adult flora become established in all infants as they are weaned and other foods are added to their diets.

Cyclical Smells

Menstruating women and their partners all know the smell associated with menstruation. The smell stems from the presence of fecal germs, which becomes problematical only during menses. Throughout the rest of the monthly cycle, the normal flora of the vaginal vault mainly comprise different strains of lactobacilli. These bacteria are acid producers, and they create a protective acid environment in the vagina (pH≅4.0) in which fecal germs such as *E. coli* cannot ordinarily survive.

When a woman menstruates, however, the blood flowing over the acidic mucosal surfaces of the vagina, as well as the surfaces of the vulva, increases the pH to a more neutral or alkaline number. This allows *E. coli* and other normal fecal flora to grow efficiently in large numbers in the vaginal vault. The blood also provides food for fecal bacteria. In fact, the smell that accompanies menses results when the bacteria break down blood and other tissue debris, thereby creating a malodorous by-product. Interestingly, prepubescent girls and postmenopausal women can carry normal fecal flora in their vaginal vaults without a bad smell, because their normal vaginal flora create a different environment than the flora of menstruating women.

The smell of menstruation aside, many people wonder if sexual intercourse during menstruation brings any increased health risk. So long as both partners are free of infectious diseases, especially hepatitis and HIV, there is generally no significantly increased risk. The exception is women who are prone to urinary-tract infections (UTIs): These women should never engage in sex during their periods. The reason is the same natural process that produces the smell of menses by encouraging the growth of *E. coli* in the vaginal vault.

There are many different types of *E. coli*. Although some are nonpathogenic, others are aggressive uropathogens. That is, they have superior adherence affinity to, and can easily infect, the urinary tract. If one of the uropathogenic strains is present in a genetically susceptible woman, sexual intercourse can introduce the *E. coli* into the bladder via the urethra. Although this can happen at any time that the dangerous strain of *E. coli* is present, menstruation will greatly increase the quantity of *E. coli* and thus the likelihood of an infection. This is compounded during intercourse because *E. coli* can be introduced into the urethra by the mechanical milking action of coitus. The normal flora of the vagina, in which *E. coli* cannot usually survive, can also be upset by the use of feminine products such as douches and deodorants. Women should use douches sparingly, and use only mildly acidic preparations, such as those that contain acetic acid (vinegar) or lactic acid. They should never use a douche that contains deodorants or germicides, as these will upset the normal flora. Interestingly, it has been shown that the use of spermicide nonoxynol-9 inhibits the growth of lactobacilli, thereby favoring colonization of the vaginal vault by *E. coli*.

As the story of the food lithe demonstrates, scientists have yet to solve some of the intricacies of the germ factory known as the human body. But we have certainly learned enough to respect and admire the delicate balances in our flora that keep us healthy from birth through old age. We can also often use that knowledge to restore health when illness strikes, and even to prevent illnesses before they happen. With this in mind, it's time to take a closer look at how infectious illness occurs and how we can treat it when it does.

5

The Enemy Within

Barbarians at the Gate

Like every other animal, every human being is a veritable germ factory in a world of air-, plant-, soil-, and water-borne germs. Only a fraction of these germs are pathogenic, that is, disease causing, but that fraction includes a number of germs that people can encounter in their daily environments, like the flesh-eating bacteria beta-hemolytic streptococcus group A the newspaper reporter found on a New York City pay phone. It also includes a number of germs in human beings' normal flora, and in fact these germs cause the majority of infections. Although every person lives mutualistically with stable populations of germs residing at specific sites on and in the body, the migration of germs from one site to another, or an explosion of germs at their normal location, can transform those mutualist germs into parasites and lead to infectious illness. Such shifts in the human flora can result from an injury, or they can very easily be an unexpected consequence of a change in people's behavior, as we'll see in more detail later in this chapter.

With so much exposure to potentially dangerous germs, why don't people continually fall ill with one ailment after another?

Part of the answer goes back to the mid- and late-nineteenth-century breakthroughs of Louis Pasteur and Robert Koch, who established the validity of the germ theory of disease by proving that specific germs cause specific diseases. Koch laid out four steps, known as Koch's Postulates, for identifying a particular germ as the cause of a disease:

1. Find a germ that is present in every case of the disease.
2. Extract that germ from a sick animal and grow, or culture, it in a pure form in vitro (literally, "in glass") in the laboratory.
3. Test the pure germ's effect in vivo ("in life") by inoculating a healthy animal with it.
4. If the animal becomes sick, extract the same germ from it, proving that germ's involvement in the disease.

Although they were otherwise fiercely competitive scientific rivals, each hoping to make more discoveries than the other, Pasteur and Koch worked within the same basic conceptual framework. Together with their followers, they soon identified the germs that cause anthrax, rabies, cholera, pneumonia, tuberculosis, diphtheria, typhoid, tetanus, gonorrhea, and many other diseases. They also developed the first effective vaccines for some of these diseases from attenuated, or weakened, strains of the germs involved. Observations from ancient times had shown that if people survived one exposure to a disease, they often acquired full or partial immunity to subsequent exposure. It was assumed that vaccines put the same immunizing process in motion, but no one yet knew how the body's natural defenses work at the cellular level, where the fate of infectious germs is ultimately decided.

Identifying the guilty germ that causes a disease opened the door to learning more about the immune system, however, because it made it theoretically possible to map every stage of disease and recovery in detail from start to finish. A continuing series of innovative laboratory devices and methods, from the electron microscope to the development of advanced molecular, genetic, and biologic techniques, has been making a reality of that dream ever since Pasteur, Koch, and their contemporaries established the principles of microbiology and laid the foundations for all subsequent medical research.

Among the many immune mysteries that science has strived to explain is the fact that contact with a disease-causing germ does not guarantee that a person will become ill. After all, even the deadliest epidemics generally leave some people who are exposed to dangerous germs unscathed, while others suffer symptoms and recover. To take a less extreme example, many healthy people must have come into contact with the flesh-eating bacteria on the Madison Avenue pay phone without getting sick. Some people simply have hardier immune systems, or possess greater tolerance for certain substances, but there is as yet no way to predict or quantify these individual differences.

To predict whether given germs are likely to cause disease, it turns out that we need to know their virulence, how many of the germs are present, and the environmental factors that will affect their survival and reproduction. Medical science uses the words "virulence" and "pathogenicity" interchangeably to describe the extent to which one thing is poisonous to another; in the case of infectious germs, this means their ability to invade a body site and emit toxins that can immobilize the body's natural defenses. To the extent that there is a distinction between the two terms, virulence implies a quantitative measure of disease potential, and pathogenicity a qualitative one. Germs of lower virulence usually must be present in large numbers to create an infection. For example, it takes at least ten thousand *Salmonella* germs to cause food poisoning. By contrast, it takes only ten to a hundred one-celled *Campylobacter* germs to have the same effect. For some highly virulent diseases, such as tuberculosis, a single germ cell can be enough to mount a deadly assault on the body. In truth, almost any microorganism, even an innocuous non-pathogen, can cause disease if it is present in large enough numbers in a favorable enough environment.

In the last chapter I mentioned that the prevailing conditions in an ecosystem can decisively influence a germ's path along a chain of infection from one host, or from a natural reservoir, to a new host. Food poisoning from *Staphylococcus aureus* is a bigger problem at summer picnics than at fall tailgate parties, because the heat of summer encourages the growth of the germs and the production of larger quantities of toxin. The same thing is true of the prevailing conditions within the environment of a host organism. If the environment is hostile enough, an invading germ won't be able to gain a foothold. But if it is friendly enough, disease may occur.

In 1934, a researcher named Theobald Smith brought these factors together in an equation, $D = N \times V/H$, where D is disease, N the number of germs, V their virulence, and H the host environment. Smith's equation remains an important tool for investigating disease processes. Its great value is that it puts the focus squarely on the interaction of germs and host, and thus in turn on the physical, chemical, and biological aspects of the host environment. The host environment really comprises two separate components: The first, or biological, component is the body's natural defenses, and the second is the physical-chemical environment at a site in or on the body.

One of the most interesting things about the human body as a host environment is that the germs in the normal flora are often closely related, or even identical, to germs that cause disease. Various staphylococci and streptococci, for example, cause boils and abscesses, food poisoning, osteomyelitis, meningitis, pneumonia, rheumatic fever, and scarlet fever, among other diseases. Yet *Streptococcus salivarius, St. mutans, St. sanguis,* and other varieties of *St. viridans* are all common mouth germs, and many people even harbor mutualistic populations of *St. pneumoniae,* which can directly cause pneumonia. Group B strep can be found in vaginal fluid, and group D strep in feces. Likewise *Staphylococcus aureus* and *S. epidermidis* frequently form part of people's normal skin flora. Yet in large enough numbers elsewhere in the body, the same *S. aureus* could lead to a fatal infection.

To understand how such germs came to live peacefully in and on the human body, it may help to consider the matter from the germs' point of view, or, perhaps even better, from the point of view of alien scientists studying the relationship of germs and other creatures on Earth. For germs that can live in or on mammals, human beings make attractive hosts because they are such a successful species. They live practically everywhere on the planet's surface, their numbers keep increasing, and they have relatively long life spans compared with many other animals. Over evolutionary time, the competition for such prime real estate inevitably favored milder strains and species of germs, or those which can exist mutualistically as well as parasitically. As I pointed out in Chapter 3, it's a better strategy for a germ to be a mutualist than a parasite, because the parasite may put itself out of house and home by killing its host too soon. Thus over countless generations of evolution the potentially lethal toxoplasmosis parasite has adapted so well to the environment of the human body that it now mainly becomes dangerous only in cases of immunosuppression. It infects at least two billion people worldwide at any one time, and most people who have it are never troubled by it.

Whether a germ is relatively virulent or nonvirulent, its viability in the human host depends above all on its precise location. Stable germ populations live mainly on the skin, in the respiratory and gastrointestinal tracts, and in the genital area, especially in the vagina. In other words, they live in what we might call the body's border zones and gateways, on the surface of the body or in areas that are directly or indirectly open to the surface and thus to the foreign territory of the outside world. They earn their keep actively by participating in essential functions, like the *E. coli* that live in our guts and help us digest the food we eat or the skin flora that eat dead skin cells and patrol for foreign germs. Or they protect us passively, because their presence denies lodging space and food to other germs.

Invading germs that arrive in the body's border areas will usually find them well garrisoned, and their ability to survive the competition for resources in those microenvironments will depend on how strong their combined numbers and virulence are, just as the Smith equation describes. For example, the vaginal flora normally include acid-producing lactobacilli, which promote the growth of acid-loving (acidophilic) bacteria. Together these microbes create a climate that is inhospitable to pathogens such as gonococcus, which causes gonorrhea. If a healthy woman engages in sex with a partner who has gonorrhea, she has a fifty percent chance of escaping infection on the first encounter. The acidic condition of the vagina and the presence of stable populations of normal flora will make it hard for the gonococci to find a place to settle, feed, and reproduce. But the woman's risks will escalate with each subsequent encounter. In general, higher concentrations and repeated challenges increase the likelihood that pathogens will be able to gain a foothold in a new environment and alter it to suit their own growth and reproduction.

In this light, we might say that the germs in human beings' normal flora serve their hosts just as mercenary soldiers once served the Roman empire. Imperial Rome filled its legions with barbarian mercenaries, who were mainly stationed at border outposts and whose chief function was to keep out other barbarians. This arrangement worked well enough on the whole, except that the barbarian gatekeepers could all too easily decide to rampage through the provinces they were supposed to protect, just as the normal human flora can often be the cause of disease.

The Roman empire declined and fell in large part because it depended too much on barbarian mercenaries and let its inner defenses atrophy. There was no vital citizen army within the body of the empire. Eventually the barbarians' deadly incursions became too much to resist, and the empire collapsed into a jumble of its former parts. The healthy human body doesn't make the same mistake, however. Instead it maintains an extraordinary array of inner defenses on full alert status. The prime operating directive for these defenses can be summed up in three words, Self vs. Nonself.

Self vs. Nonself

Anyone who spent, I won't say squandered, childhood hours reading *Mad* magazine will recall a continuing comic strip feature titled "Spy vs. Spy." In every issue two opposing spies tried to outwit each other with their fiendishly clever schemes, but any victories were sure to be only temporary. The two agents were drawn identically in every detail and could not be told apart, except that one was dressed in white and the other in black.

A similar, if much more serious, contest goes on in the human body between Self and Nonself. Once an infectious germ or other foreign agent penetrates the body's outer defenses of skin and mutualistic flora and infiltrates the bloodstream and deep tissues, it might well congratulate itself on being home free. The blood's proteins offer a rich food source, and the bloodstream can carry a germ's fast-growing population of clones anywhere in the body. Luckily for us human beings, there is one snag that can foil the enemy germ's plans. Every antigen, as foreign agents in the body are known, has a protein marker on its cell membrane that identifies it to the body as Nonself. The body can even recognize the cells of another individual of the same species as Nonself, a fact that greatly complicates organ and tissue transplants.

When an antigen's presence is announced by its Nonself protein marker, the body's internal defense swings into action with a devastating one-two punch employing lymphocytes, white blood cells created by the lymph system. Lymphocytes known as B cells (because they originate in bone marrow) produce protein molecules called antibodies. The antibodies relentlessly hunt down antigens and bind with them, preventing them from attaching to host cells and beginning a cycle of growth and reproduction. Then helper T cells and cytotoxic T cells (both originating in the thymus) can join in with what are called cell-mediated immune responses (CMI). Helper T cells release soluble proteins that make it easier for cytotoxic, or cell-killing, T cells to engulf and destroy the antigens. In addition, the bloodstream contains a smaller group of lymphocytes labeled NK, or natural killer, cells, which kill tumor cells and virus-infected cells. The NK cells (sounds like a casting call for an Oliver Stone film, doesn't it?) and cytotoxic T cells are also both known as phagocytes (literally, "cell eaters"), because of how they gobble up antigens; the whole process is known as phagocytosis.

The immune response doesn't stop there. After knocking out the enemy germs, some of the B cells and T cells become long-lived memory cells. Their memory is for the specific Nonself protein marker on the germs they just killed. This cellular memory makes it possible for the body to deliver an even more remarkable one-two punch. If the same, or even a closely related, strain of germs should ever threaten the body again, the memory B and T cells will be able to cut the body's response time dramatically. This combination of primary and secondary responses is the key to the mystery of immunity. A first exposure to a pathogen, or to a vaccine that employs the pathogen or a close germ relative in weakened form, sparks the formation of memory cells that will instantly zero in on that pathogen if a second exposure occurs.

How does the immune system tell enemy germs from benign, mutualistic ones, initiating an inflammatory response that destroys hostile invaders while leaving friendly bugs unscathed? A recent scientific discovery suggests that a special protein of the Nod 2 gene enables white blood cells called monocytes to distinguish between the normal gut flora, which they ignore, and invading bacterial pathogens, which they engulf and destroy. Nod 2 can activate a chemical—nuclear factor kappa B (NF-kB)—that helps regulate this process. Inhibitory kappa B (I-kB) ordinarily suppresses NF-kB. But when pathogens are present, their activity interferes with I-kB and allows NF-kB to be released.

All in all, then, human beings are well protected from the skin in. The skin itself is an effective barrier against the majority of pathogens, but the body doesn't leave this outer wall unguarded. The mutualist germs in our skin flora patrol the ramparts. Other mutualist germs help keep the gateway areas, the respiratory and gastrointestinal tracts and the genitals, healthy. And over the course of the last century and a half, medical science has developed a strong backup army of antibiotics, antitoxins, and vaccines to call on whenever the immune system needs help.

The odds are clearly in human beings' favor, but even the most favorable odds don't guarantee a victory in every case. Sometimes the body overreacts to a foreign substance, and in the process harms itself. This hypersensitivity is the cause of allergies and can even lead to tissue damage, anaphylactic shock, and death. The other danger in the body's regulation of the immune system arises when, for reasons yet unknown, the body mistakes its own normal, healthy cells for foreign or abnormal ones. When the body develops antibodies and activates CMI against itself, autoimmune disorders such as lupus, myasthenia gravis, diabetes, and Graves' disease can result.

Moreover, no immune system is bulletproof. Over the four billion or more years of their existence on Earth, germs have evolved all sorts of strategies for infiltrating host organisms, including human beings. Consider the gonococcus which causes gonorrhea. Before this bacterium can initiate disease in a human being, it must first adhere to human body cells. The gonococcus accomplishes this task by attaching to our cells with its pili, hairlike protein filaments that arise from its cell wall. The body responds by producing antibodies against the gonococcus germs. But when the gonococci find themselves under attack, they switch production of pili proteins from one type to another. This baffles the antibodies, because the body doesn't recognize the new pili even if it's been infected with gonorrhea before. To complicate matters, gonococci can switch in turn to as many as three hundred different varieties of pili; one or two variations will usually be enough to thwart the body's immune system.

Another brilliant strategy germs employ is to cover themselves in capsules. An encapsulated strain of *Streptococcus pneumoniae* will have a smooth appearance, and the phagocytes in the bloodstream will not be able to engulf and destroy these "S" types. In contrast rough, or "R," strains have lost their virulence, have no capsules, and are easily detected and eaten by the phagocytes.

Many different species of germs have independently developed an ability to survive phagocytosis. After they engulf enemy germs, phagocytes release potent enzymes that complete their destruction. But germs such as the one that causes tuberculosis *(Mycobacterium tuberculosis)* can actually prevent these killer enzymes from operating, and thus use the phagocytes to spread themselves throughout the body. *Salmonella, Listeria,* and *Legionella* germs also have this capacity to turn the body's defenses against itself.

Germs also produce their own toxic enzymes, which can incapacitate the body's defenses. For instance, the cells in the body are held together by hyaluronic acid, which acts as a sort of cement. Some germs can release hyaluronidase, which destroys hyaluronic acid, allowing for more efficient invasion of body tissues. *Staphylococcus aureus,* which can easily contaminate foods like potato salad or egg salad, releases enterotoxins that cause diarrhea. And many other toxins are produced by other sorts of germs.

Sometimes an infectious agent can't get into the body on its own, but can hijack a friendly germ, converting it into a veritable Trojan horse that is welcomed at all its usual sites in the body, where it then sows disease. Recent food-poisoning scares involving apple juice and alfalfa sprouts have been linked to such infiltration of *E. coli,* a digestion-aiding germ that is part of our normal mutualistic flora. In relatively recent times a strain of *E. coli* designated 0157:H7 acquired a very potent toxin, Shiga toxin, which originates in *Shigella* bacteria. Shiga toxin causes dysentery, a severe, bloody gastroenteritis that can be life threatening. It is generally accepted that a bacterial virus called a bacteriophage was responsible for planting Shiga toxin in this particular strain of *E. coli.* Thus viruses can infect a lowly fellow germ, as well as a human being, and endow it with potentially lethal capabilities.

Another interesting germ strategy is exemplified by the parasite *Trypanosoma brucei,* which causes African sleeping sickness. This disease attacks in recurring waves. At one point you're well, next moment you're sick, then you feel well again until the next wave strikes, and this process can go on and on. The disease recurs because the trypanosomes can change the protein marker on their cell membranes. The body identifies the trypanosomes as Nonself by reading this protein marker, and then produces antibodies against them. In response, the germs can switch the protein, a glycoprotein known as variant specific glycoprotein (VSG), into more than one hundred different forms in the course of a single infection. The result confuses and overwhelms the immune system just when it is on the verge of killing off all the intruders.

One last germ strategy, among many others, deserves a special mention. HIV causes AIDS by infecting the body's T cells, the immune system's frontline soldiers. The infection of T cells cripples the immune system and makes a person with HIV vulnerable to infection by a host of germs that a person with a full complement of T cells could shrug off with little or no distress. Later on, I'll discuss how the latest treatments for AIDS work to preserve T cells and keep them battle ready to fight infections.

The never-ending struggle of Self vs. Nonself in the human body is from one perspective a story of coadaptation by human beings and germs. Over evolutionary time, parasitic germs will gradually become mutualistic, or at least somewhat less threatening to their hosts, simply because the milder strains will be able to live longer, reproduce more, and spread to more hosts than the stronger strains. Remember, if the host dies before its time, the parasite is out of food and shelter. So as I mentioned above, even HIV will likely develop less lethal strains as it adapts to living in human beings. Likewise, medical science will continue to discover more about germs and their strategies, and this will give us more tools to combat the dangerous ones and keep us healthy. At least, it will if we exercise common sense and remember that our innovations sometimes have unwanted consequences.

"We have met the enemy, and he is us."

In the old comic strip *Pogo,* a character memorably summed up the human penchant for self-sabotage by remarking, "We have met the enemy, and he is us." I sometimes think that line should be added to the Hippocratic Oath as a reminder of how people can undermine their health, not just with reckless behavior like smoking, but also with practices specifically designed to promote health and hygiene.

Modern medicine has had no greater triumph than the discovery and development of antibiotics, substances that kill dangerous germ cells but mainly leave the cells of the body unharmed. Although antibiotics can disrupt the body's normal flora and cause gastrointestinal upset, for example, and although some people are allergic to specific antibiotics such as penicillin, for the most part we can use them without any worry that they will harm us.

But there can be too much of a good thing. The ease with which antibiotics knock out disease-causing germs has led to overuse that, at least from the point of view of the population as a whole, amounts to a form of unintended self-abuse. The problem is that where antibiotics are present, natural selection will favor the growth of antibiotic-resistant germs. Consequently, the most fiercely antibiotic-resistant germs will thrive in hospitals, where antibiotic use is heaviest. Because of antibiotic-resistant germs inside and outside hospitals, tuberculosis, for which medical science once thought it had a sure cure, is reemerging as a public-health menace, raising the specter that the disease could again become a leading cause of death among adults. Similar problems are arising with other diseases, such as gonorrhea.

Later in the book I'll talk about some of the potential solutions to antibiotic resistance. Here I want to emphasize that just as a whole society can unknowingly make itself more, rather than less, vulnerable to infectious germs by relying too heavily on antibiotics, so individuals can often change the physical, chemical, and biological conditions in and on their bodies for the worse without realizing it. On a relatively innocuous level there are all the creams, cosmetics, deodorants, and perfumes we apply to our bodies, sometimes changing the conditions of a body site so much that a rash or other undesired reaction occurs. On a much more serious level, there is the story of toxic shock syndrome, which exemplifies how disregarding human beings' delicately balanced microbial ecology can lead to disease and death.

Toxic shock syndrome (TSS) is a multisystem disease, meaning it can affect more than one organ system in the body. The symp-

toms mimic those of scarlet fever and range from a mild, flulike illness to heart, kidney, liver, and respiratory failure, including fever
equal to or greater than 102 degrees Fahrenheit, skin eruptions
with red rash and subsequent peeling of the skin, nausea, vomiting, diarrhea, low blood pressure, and dizziness. The disease is
caused by exotoxins produced by strains of *Staphylococcus aureus,*
which many people have as part of their normal flora but which
can prove fatal if its toxin gets into the bloodstream through a
wound infection, a complication of influenza, or some other
means. The *S. aureus* toxin most directly implicated in the disease
has been dubbed toxic shock syndrome toxin-1 (TSST-1).

Toxic shock syndrome burst into public attention in 1980, and I
became actively involved in research into it at that time. By 1983
more than 2,200 cases had been reported to the Centers for Disease
Control in Atlanta. A large proportion of these cases were fatal and
followed the same pattern. A typical case, the subject of one of the
first lawsuits over toxic shock syndrome, was that of Patricia Kehm
of Cedar Rapids, Iowa. In September 1980, Pat Kehm was a
twenty-five-year-old, historically healthy mother of two young
daughters. In the early morning hours of September 5, 1980, she
woke up with a burning fever and chills. She soon began vomiting
and having diarrhea. Meanwhile, her temperature reached 103
degrees Fahrenheit. Pat thought she was coming down with the flu.
Later in the day Pat's husband, Mike, became concerned that her
condition was deteriorating further and took her to the local hospital's emergency room. The admitting doctor catalogued her symptoms as sunburn-like rash, red flushing, severe dizziness, low blood
pressure, and sore throat. Her chest and other parts of her body
showed differences in color, and her legs and arms were starting to
turn bluish. The admitting doctor initially misdiagnosed the case as
strep sore throat and administered penicillin, an antibiotic that is
generally not effective against *Staphylococcus aureus.*

Pat Kehm soon went into shock, and she was given IV solutions and drugs in an effort to boost her blood pressure. At this
point a nurse noticed that she was wearing a tampon and
removed the bloodied pledge. Shortly thereafter Pat Kehm's heart
stopped; when efforts to resuscitate her failed, she was pronounced dead.

As research into toxic shock syndrome progressed, it became clear that the disease was not really new. In 1978, Dr. James Todd, a Denver pediatrician, described an unusual illness in four girls and three boys between the ages of eight and seventeen, the symptoms of which were very like those Pat Kehm later suffered and which he named toxic shock syndrome. (The disease is not contagious; the children's cases were connected only in that they all occurred around the same time.) Five of the seven patients were found to have an exotoxin producing strain of *S. aureus* in the nose, throat, vagina, or other body sites. Todd surmised correctly that the *S. aureus* was producing toxins that were released into the bloodstream. The first case to be diagnosed and treated, what epidemiologists call the index case, was that of a fifteen-year-old-girl who had this *S. aureus* strain in her throat and vagina. Later it emerged that she had been menstruating and wearing a tampon when she became ill.

Fifty years before the Denver cases, in 1928, a pediatrician in the small city of Bundaberg in Queensland, Australia, inoculated twenty-one children, including his own son, with a vial of diphtheria antitoxin contaminated with *S. aureus*. Within twenty-four hours, eighteen of the twenty-one children became ill, and eleven children died. A twelfth child, the final fatality, died at the beginning of the next day. A Royal Commission investigated the disaster and reported the same scarlet-fever-like symptoms that would later occur in Denver. The nine children who survived all developed abscesses where they had been injected with the antitoxin, and these abscesses were found to contain *S. aureus*.

As early as 1908, the medical and scientific literature reported that numerous strains of *S. aureus* caused an illness resembling scarlet fever. In fact, the disease may go back much further than that. Around 430 B.C., during the Peloponnesian War between Athens and Sparta, an epidemic raged through the Athenian army, killing tens of thousands of men, including Pericles, the leader of Athens. The epidemic so weakened the Athenian side that it lost the war to the Spartans. In the past this plague was attributed to smallpox or an assortment of other microbes. But in the light of the toxic shock syndrome cases of the 1980s, medical detectives at the Centers for Disease Control in Atlanta were inspired to examine the great Greek historian Thucydides' descriptions of the disease. Thucydides served on the Athenian side in the Peloponnesian War, and his eyewitness account has always been prized for its accuracy and detail. From the symptoms that Thucydides catalogued, the CDC researchers concluded that the epidemic was caused by viral influenza complicated by toxic shock syndrome.

If toxic shock syndrome was not a new disease, the question remained of why its incidence increased so dramatically around 1980. Since the *S. aureus* that causes the disease has probably been part of the normal human skin and vaginal flora since our species' earliest beginnings, one possibility that needed to be considered was whether the germ or its toxins had evolved into a more virulent form. Studies showed, however, that *S. aureus* has been able to produce TSST-1 since at least 1900. Other studies showed that antibodies to TSST-1 are historically equally distributed among the male and female members of the population. Not all people have *S. aureus* in their skin flora and not all women have the germ in their normal vaginal flora, but many do.

What was most interesting here was the discovery that the incidence of antibodies to TSST-1 is also historically age-related. Testing of stored blood samples from the 1960s, 1970s, and 1980s revealed that only about half of people under sixteen years of age have the antibodies, whereas eighty percent of twenty-year-olds have it, and more than ninety percent of those thirty-years old or older have it. Given how common *S. aureus* is, it makes sense that as people grow older they are more and more likely to be exposed to the germ in some way and to develop antibodies against its toxins.

This explained why most of the victims of toxic shock syndrome in the late 1970s and early 1980s were young people. But it also suggested that the disease should be about equally likely to strike young men and boys as young women and girls. In fact the disease seemed to be decidedly gender biased. Of the more than 2,200 cases reported to the CDC through June 1983, ninety percent were associated with women who were menstruating at the time they became ill. Most of these women were young, and ninety-nine percent of them were using tampons.

The Tampon Connection

In regard to TSS, then, the central question facing medical science at this time was whether the cause of the epidemic was the germ, the women, or the tampons. As we've just seen, the germ and its toxins had not changed. It was, of course, possible that the women who became sick had health histories or behavioral habits that suppressed their immune systems and made them susceptible to toxic shock syndrome. But although the tampon makers searched hard for such correlations, none emerged. That left the tampons.

The connection of toxic shock syndrome and tampons was noted early on. By midsummer 1980, the CDC had firmly established a link between the disease and the device and warned physicians to be alert to it. Later that year, the American College of Obstetricians and Gynecologists also issued an advisory on tampon use and TSS, and the next year the Institute of Medicine of the National Academy of Sciences followed suit. It was clear that tampons were somehow making women vulnerable to TSS, but exactly how they were doing so was a baffling question. After all, women have been using tampons since 1933, when Dr. Earl Haas invented them so that his wife, a ballet dancer, could perform even on "those days."

Menses occurs approximately every twenty-eight days, the length of the lunar month. The menstrual cycle begins when a mature egg is released into the uterus and the lining of the uterus thickens to provide a good environment for a fertilized egg to develop into a fetus. If the egg is not fertilized, however, the uterus sheds its thick lining in the menstrual flow, which drains through the vagina. For hygienic and other reasons, women have been absorbing their menstrual flow in one way or another since the earliest times. For example, the ancient Egyptians used papyrus paper pads to collect menses. In traditional African, South Pacific, and South American Indian cultures, women commonly used grasses. The Southeast Alaskan Tlingit and Haida people used a variety of mosses to control their flow. Much later, in Western societies, women employed reusable cloth rags, which were cut up into strips. (From that practice, by the way, came the phrase "on the rag.") Soon after the turn of the twentieth century, disposable cotton pads became the predominant menstrual aid, at least for those women who could afford them, until Dr. Haas devised his internally worn, all-cotton tampon.

Essential as the menstrual cycle is for the continuation of the human species, it can be quite an inconvenience for women. But more than that, it can also constitute a health hazard, albeit usually a rather mild one. Except during menses, the vagina is an oxygen-free, highly acidic environment, owing to the action of women's normal vaginal flora. As I mentioned above, this makes the vagina generally inhospitable to invasive germs, including ones as virulent as the gonococcus that causes gonorrhea. The menstrual flow neutralizes the vagina's normally acid environment, rendering it much more susceptible to infectious germs. Interestingly, this may have been a less serious issue for women's health in preindustrial times. Some researchers estimate that because of frequent pregnancies and breast-feeding, women in preindustrial societies menstruated only about a third as often as women in industrial societies.

In modern societies, Dr. Haas's tampon was a godsend. It represented the most efficient and hygienic means yet of collecting the menstrual flow and letting women get on with their lives at the same time. Women adopted it eagerly with no increase in the incidence of toxic shock syndrome. But for many women who preferred internal tampons over pads, the all-cotton tampon was still not a perfect answer to heavy menstrual flows that could bypass a sodden tampon and spot one's clothing. In the 1970s, a number of companies tried to satisfy women's desire for more absorbent tampons by manufacturing them with synthetic materials or mixtures of synthetics and cotton rather than cotton alone. The company that took the lead in this was Procter & Gamble, which had never made tampons before. P&G introduced a superabsorbent synthetic tampon named Rely and quickly gained a sizable share of the market, as much as twenty-five percent in some areas of the United States.

Not only did the synthetic fibers in the Rely tampon absorb more menstrual fluid than cotton, the materials also expanded to fill more of the vaginal space and act as a dam until the tampon was removed. This too helped to cut down on menstrual bypass and spotting. As the product packaging put it, Rely "even absorbs the worry." To compete against Rely and keep from losing still more of their market share to an upstart product, the established tampon manufacturers quickly followed P&G in offering tampons made primarily with one or more synthetic fibers such as carboxymethylcellulose, polyester, polyacrylate rayon, and viscose rayon, sometimes mixed with cotton and sometimes not. The result was that from 1980 until 1993, no all-cotton tampon was commonly available in the U.S. market.

The health advisories that the CDC and other medical authorities issued in 1980 and 1981 took special note of this change in the composition and structure of tampons. The CDC even singled out Rely tampons. At a twenty-five percent market share, P&G's new product was responsible for seventy-five percent of the TSS cases. Patricia Kehm was wearing a Rely tampon when she became ill. On September 22, 1980, in a stopgap public-relations and legal move, P&G voluntarily took Rely off store shelves, aware that the Food and Drug Administration was just about to order the product withdrawn from the market.

By this point the connection between TSS and synthetic tampons seemed incontrovertible to everyone but the manufacturers. The companies were waging an all-out battle to protect their product lines and their profits in the court of public opinion and in the legal courts, where victims who survived TSS and the families of those who didn't, like Pat Kehm, were soon filing product liability lawsuits. To settle the matter, it would be necessary to establish the precise action of the synthetic tampons in the vaginal environment. Here is where I was able to make a contribution, together with my colleague Dr. Bruce Hanna.

It was actually my wife, Josephine, who got me started on the TSS problem. As the crisis became a big news story, Josephine asked, "Am I in danger of getting toxic shock syndrome? I use superabsorbent tampons." She was not using Rely, but another brand. I was able to assure her that she was not in danger, because she had no *S. aureus* in her normal flora. But her question set me thinking. Like many other households all across the country, we had received a consumer sample of Rely in the mail. I opened the package to examine the tampon and test its absorbency with water. It was amazing to see how much water the Rely tampon could hold, and how large a size it expanded to. In fact, one Rely tampon could absorb an average woman's entire menstrual flow.

Louis Pasteur famously remarked, "Chance favors a prepared mind." Although my work on TSS hardly merits mention with Pasteur's towering achievements, my education and my prior research into the ecology of human beings' microbiology did prepare me to make some lucky guesses about the disease. I knew that TSS had to arise from a constellation of physical, chemical, and biological factors, and in each category certain surmises occurred to me, as follows.

Physical factors: Beginning over fifty years ago, microbiologists found that staphylococci require the presence of agar or other thickening agents to produce toxins of enough quantity and potency to cause illness. The thicker, or more viscous, the medium in which *S. aureus* is growing, the more toxin will be released. This is why staphylococci are likeliest to cause food poisoning in rich, thick dishes like potato, macaroni, and egg salad. The menstrual flow has a relatively high viscosity because of the proteins in the blood. My little experiment at home with a Rely tampon and a glass of water suggested that in the case of women wearing synthetic tampons, the synthetic fibers in the tampons might dramatically increase viscosity in the vagina because of the great amount of blood they would absorb and because of the greater viscosity that synthetic fibers have to begin with.

A second important physical factor is surface area. The more extensive the surface area that *S. aureus* has available to grow on, the more toxins it will create. By the nature of their materials, synthetic tampons have much more surface area than all-cotton tampons.

Chemical factors: The optimal pH (acid/base proportions) for TSST-1 production is about 7. This pH level depends in part on there being enough carbon dioxide in the immediate environment. TSST-1 is produced in higher quantities in an atmosphere of ten percent carbon dioxide. During menstruation, the carbon dioxide level in the vagina rises and the pH of the vagina increases to about 7. This is because the menstrual blood, which has a pH of about 7.4, buffers the normally acidic environment of the nonmenstruating vagina, which has a pH of about 4.2. TSST-1 production has also been shown to be stimulated in an aerobic environment containing about twenty percent oxygen. The nonmenstruating vagina contains no oxygen, but inserting a tampon that swelled up with blood and created a new microenvironment inside the vagina might well help raise both carbon dioxide and oxygen levels to a point favoring the production of TSS toxins.

Biological factors: It was already well known in 1980 that *S. aureus* makes pyrogenic, or fever-inducing, toxins. In fact, a temperature of 102 degrees Fahrenheit or higher is ideal for production of TSST-1, and this is exactly the range of fever induced by toxic shock syndrome. In this regard the disease functions as a feedback loop: When TSST-1 is released into the bloodstream, it induces a fever, which in turn stimulates the production of more TSST-1.

Having thought the matter through in this way, I wrote a letter to Procter & Gamble on October 10, 1980, with copies to the CDC and the Food and Drug Administration, laying out what I suspected the logical progression of the toxic-shock disease process to be. "Imagine hundreds of . . . 'toxin factories' or 'toxin chambers' in every super absorbent tampon," I wrote in part. "The super absorbency of the product brings in enormous amounts of nutrients to insure this phenomenon."

The research that Dr. Bruce Hanna and I conducted at New York University Medical Center eventually confirmed my surmises in every detail. Exploring the tendency of tampons to stimulate *S. aureus* toxins associated with TSS, Dr. Hanna and I found that the greatest increases occurred in tampons made with polyacrylates and in Rely tampons, which were made with carboxymethylcellulose and polyester. Lesser amounts of TSS toxin were associated with blended rayon and cotton, and there was no increase, compared with controls, in tampons made solely with cotton. We published these results in the *American Journal of Obstetrics and Gynecology* in mid-1985. With this article and one in the previous issue by Dr. Gorm Wagner and his colleagues at the University of Copenhagen, who found that inserting a tampon into the vagina increases the carbon dioxide and oxygen levels to a point favorable to TSS toxins, the tampon-TSS connection was scientifically explained, a fact that many subsequent studies have confirmed.

Dr. Hanna and I have continued to investigate TSS and tampons. In a recent article we reported the results of a comparative analysis of a variety of tampons, vaginal contraceptive sponges, nasal surgical tampons, gauze packs, vaginal cups, vaginal pads, diapers, surgical drapes, gelatin foam surgical sponges, and surgical cotton. We compared these substances and products for their propensity to induce TSST-1 production when brought into contact with liquid broth cultures containing TSS-causing strains of *S. aureus*. Our data revealed that the amount of TSST-1 was greatest in the liquid broth cultures grown in the presence of all synthetic tampons but only some other products. TSST-1 was either absent, or present only in trace amounts, in liquid broth cultures grown in the presence of cotton controls. This showed not only that the physical and chemical environment for TSS toxin production was best in synthetic tampons, but that the problem extended to a variety of other man-made health and surgical products.

There were no all-cotton tampons on the U.S. market when we conducted our original tests, so I created mock all-cotton tampons in the laboratory to serve as our control substances. When all-cotton tampons finally came back on the U.S. market, in 1993, I reran all the experiments. The retests confirmed that synthetic tampons dramatically favored toxin production, whereas there was no toxin production in the all-cotton variety.

To sum up, synthetic tampons, with their superabsorbent materials and greater surface area, create a microenvironment within the vagina that constitutes an ideal breeding ground for any toxin-producing *S. aureus* that a woman may have in her normal vaginal flora or harbor there transiently. It is as if a woman decided to give the toxin-producing *S. aureus* a fortified garrison, in the form of the synthetic tampon, in which to build up its strength and numbers before flooding the bloodstream with toxins. This effectively turns a mutualist germ into a viciously parasitic one.

This also explains one of the most frightening aspects of toxic shock syndrome, which is that surviving it does not confer immunity, as would be usual in most diseases. The reason is that the TSS toxins suppress the body's lymphocytes and keep the B cells from transforming into plasma cells that can make antibodies to fight the toxins. The situation is even more dire in tampon-related toxic shock syndrome: The tampon is such a fertile breeding ground for *S. aureus* that it leads to toxin production far greater than would be found in any naturally occurring incidence of the disease. Humanity sometimes likes to flatter itself that it can improve on nature. Here's a case where humanity did improve on nature, by making a potentially deadly disease even deadlier.

It is possible to acquire sufficient antibodies against TSS toxins. But this can only happen incrementally, in small steps. As we grow and mature and become exposed to *S. aureus,* our immune systems may have many occasions to respond to minuscule quantities of TSST-1 and other TSS toxins. Over time, the antibodies can become numerous enough to safeguard a person against TSS. This was one of the implications of the data acquired when the stored blood samples from the 1960s to 1980s were tested, as I mentioned above.

If a woman does not have *S. aureus* in her vaginal flora, transiently or otherwise, she has no need to worry about tampon-related toxic shock syndrome. But very few women have the means to learn whether or not they carry *S. aureus,* as my wife was able to do. Women who have *S. aureus* in their flora, but who have sufficient antibodies against TSST-1, will also be at lower risk for TSS. But again, few women have the means to find this out about themselves. Among the complications is that not every strain of *S. aureus* produces TSS toxins. On the whole, now as in the 1980s, women are really at the mercy of the manufacturers and of regulatory agencies that have limited resources for testing the safety of products independently of what the manufacturers themselves do.

The fact is that women are still getting TSS from wearing tampons containing synthetic materials. Most tampons available in the United States continue to be made with viscose rayon, sometimes mixed with cotton and sometimes not. Although less absorbent than the carboxymethylcellulose, polyacrylate, and polyester tampons that were withdrawn, viscose rayon tampons are more absorbent than all-cotton ones. They can be a contributing cause of TSS in a woman who harbors a TSS-toxin-producing strain of *S. aureus,* either transiently or as part of her normal flora, and who has not developed sufficient antibodies to defeat the toxins. Viscose rayon can be made from sawdust and it is a very inexpensive material compared with cotton, which is less readily available and thus more expensive because of variations in crop production and so on. The bottom line is that the manufacturers apparently find it more profitable to make tampons from sawdust, and to pay any incidental liability costs that might arise, than to make them from cotton.

In my opinion, the number of current cases of TSS is underreported, perhaps significantly so. In 1989 only forty-five cases of menstrual TSS were reported to the CDC. But to be reported to the CDC, an incidence of TSS must meet what is called a strict case definition. This requires that the patient have symptoms in all of five major categories: fever of 102 degrees Fahrenheit or greater, rash, skin peeling one to two weeks after the illness begins (assuming, of course, that death has not already occurred), low blood pressure or dizziness, and involvement of three or more organ systems. If symptoms are absent in one or two categories but present in the others—for example, if a woman has a fever of 101.9 degrees Fahrenheit rather than 102, or if she does not have a rash—the situation may fulfill a clinical diagnosis of TSS. For every case I see that meets the CDC's strict case criteria, I see five or more that fit the clinical bill. With that in view, here is a **Protective Response Strategy** for using tampons safely.

HOW TO USE TAMPONS SAFELY

- **If you have ever contracted TSS, never use tampons again!**

- **Use only the absorbency necessary to control flow.**

- **Leave a tampon in place no longer than six to eight hours, depending on menstrual flow.**

- **Whenever possible, alternate between the use of tampons and pads.**

- **Do not sleep with a tampon in place. Use a pad instead.**

- **Never use a tampon for any reason other than menstruation, except as directed by your doctor.**

- **Use one-hundred-percent cotton tampons whenever possible, as they provide the lowest risk of contracting TSS.**

- **Know the symptoms of TSS. At the first sign of these symptoms, remove the tampon and rush to your doctor. Be sure to tell the doctor that you are menstruating and that you have been using tampons.**

- **Remember that TSS still occurs in women who use tampons. Hence there is a risk, albeit a small one, of contracting TSS by using any tampon.**

One of the most interesting aspects of the TSS epidemic was the tampon makers' insistence that they were only giving women the superabsorbency they wanted. It's certainly true that women wanted more absorbent tampons, and still prefer them, all other things being equal. But with any product we buy to use in or on our bodies, our primary interest as customers is in staying safe, and that need above all is what we want companies to respect and serve. Unfortunately, in my opinion the manufacturers put their superabsorbent tampons on the market without doing the appropriate qualitative or quantitative studies needed to determine the microbiological impact of inserting such synthetic chemical compounds in the human body. Ralph Nader and other "green" politicians often speak of big companies as parasites on the people. But in regard to toxic shock syndrome, the companies making tampons could have learned a lot from studying parasitic germs and how they adapt as fast as possible to being good mutualist residents in their host organisms, especially when the host, in this case the man-made host combination of the menstruating vagina with a blood-soaked synthetic tampon inside it, provides an ideal physical and chemical environment for the production of deadly toxins.

This suggests that women, and all those who are concerned about women's health, should lobby for research and development of menstrual aids that work with the normal vaginal flora to promote health and hygiene. To that end, in 1997 I was glad to be able to advise New York Congresswoman Carolyn B. Maloney on drafting a bill to create a Tampon Safety and Research Act, which would direct the National Institutes of Health to conduct research to determine the safety of tampons and related products. As a man I can only feel that if men rather than women used tampons, this bill would long since have become law. Instead, it has been voted down each time Congresswoman Maloney has proposed it. Luckily, she hasn't given up the fight.

Good Manufacturing Policy

In any case such R&D would be the welcome opposite of the tampon makers' strategy, which was to ignore, dismiss, and suppress reports of unwelcome side effects from using synthetic tampons. The FDA encourages pharmaceutical and other companies in the health-care industry to practice Good Manufacturing Policy (GMP). Some of the new-wave toothpastes that have recently come on the market offer an excellent example of what such policies can accomplish, simply by capitalizing on the natural tendencies of germs.

In the normal mouth flora the ratio of anaerobic to aerobic germs is about 1000:1, and these anaerobic germs can cause bad breath. Adding peroxide to toothpaste can significantly reduce this problem by releasing oxygen in the mouth environment and thus reducing the number of anaerobes. Similarly, bacteria growing on the surface of the teeth in plaque utilize the sugars in the food we eat to produce acids that demineralize the teeth. If unchecked, this process leads to the accumulation of a hard, chalky, calcium-rich substance called calculus on and in between the teeth. A dentist or dental hygienist has to remove the calculus with a pick. But a toothpaste containing baking soda, which is very alkaline, can stop the calculus before it starts, or at least significantly reduce it, by making the mouth environment less acid. In addition, a less acid environment in the mouth helps retard the formation of cavities.

Or consider this. The tooth surface is made up of a complex of calcium and phosphate salts called hydroxyapatite. This substance readily adsorbs the proteins in particular glycoproteins from the mouth; plaque is the formation of these proteins as a film on the tooth surface together with the bacteria that adhere to it. The plaque then adsorbs other salivary components and more bacteria in layers, one atop the other, in a process called epiphytic growth. Capitalizing on knowledge of how this process works, one manufacturer has created a new toothpaste containing triclosan, a chemical that retards epiphytic growth and thus reduces the formation of both plaque and calculus. This not only makes the teeth look better, it also helps prevent cavities.

A manufacturer must consider the microbiology of a body site very carefully, because anything that affects the physiochemical aspects of a body site can affect a person's health. The tampon–toxic shock connection shows what can go wrong when adequate attention is not paid to these matters, but the new toothpastes show what can go right when it is. The benefits of tailoring consumer health and hygiene products to the specific physical, chemical, and biological aspects of a body site have only begun to be exploited. But I suspect this may well be a wave of the future. The germ factory of the human body simply offers too many opportunities to profit from, in money and health, if we only remember to respect the mutualist accommodations that human beings and their resident germs have made to each other over the course of their joint history on Earth.

PART II

The Seeds of Disease

Germs may infect those who live with persons infected, and germs can be preserved for a certain time, not only in fomes but also in the air.

—Girolamo Fracastoro, 1546

6

Person to Person

Howard Hughes left a vast fortune to a medical research foundation bearing his name. But when I think of him hiding from germs and the outside world in his Las Vegas hotel suite, I wonder how accurate his own medical knowledge was. If he had the understanding of germs and their relationship with human beings that we've just surveyed, he might not have felt it so necessary to isolate himself from everyone except his private retinue of Mormon bodyguards and male nurses.

It's not that germs don't frequently threaten human health. Of course they do. And in one sense Howard Hughes's response to that threat was perfectly logical. As we saw in Chapter 3, infectious germs can spread through direct or indirect contact with another person or an animal such as a pet, by contact with a common vehicle such as contaminated food, by airborne dispersal, or by a vector such as a mosquito. In this and the next three chapters, we'll look at each of these mechanisms in turn and assess their implications for human health. But if you're going to worry about germs, the number-one thing to worry about is contact with other people. Eighty percent of all infections, from the common cold and flu to venereal diseases and Ebola fever, are transmitted by people touching one another or touching objects that others have touched. Common cold, flu, and stomach viruses, for example, can live on the fingertips for hours, and they can survive on the surfaces of objects for days. The technical name for a contaminated object that can transmit disease is fomite, pronounced "foe-might," which comes from the Latin word "fomes," meaning tinder. As the name suggests, contact with a fomite can set an infectious illness on fire.

But as we saw in the last chapter, most germ contacts can be shrugged off by a healthy immune system. Moreover, if we exercise common sense and a little self-restraint, we can prevent most dangerous germs from threatening the immune system in the first place. It is especially important to remember that, in the words of a Mayo Clinic slogan, "The ten worst sources of contagion are our fingers." We touch our eyes, noses, and mouths, usually unconsciously, more than twenty times a day. So if we are alert to possibly dangerous germ contacts from touching other people and touching fomites, and if after such contacts we can keep ourselves from touching our eyes, noses, and mouths or eating or drinking anything until we've washed our hands, we can rest easy at night and quiet most, if not all, of any paranoid fears we share with Howard Hughes.

Alertness depends on knowledge, and in this chapter I'll be talking about a number of germ risks that we encounter in our daily contacts. Some of them will surely surprise you. In each case I will offer a Protective Response Strategy that can help to keep you and your family from undue risk of infection. The same strategies can also help minimize the extent to which you make friends, family, coworkers, and others sick unnecessarily when you have a cold, the flu, or perhaps an even more serious infectious ailment. But before going any further, I want to emphasize that the number-one **Protective Response Strategy** is to wash your hands frequently as described at the end of Chapter 2. In what follows I'll be repeating that advice quite a bit in relation to specific germ threats and potential germ contacts. Let's begin where many of us assume we don't have to worry, in our own homes.

Home Truths

The scene is a typical suburban home on a weekday morning. Before the alarms start ringing, all the members of the household are safe in their beds. Or are they? If they are sleeping on so-called reconditioned mattresses, they are likely resting on things that would turn their stomachs if only they could see them. NBC's "Dateline" once asked me to examine a number of reconditioned mattresses, and I found their contents to be rife with disgusting odors, stained with blood and urine, and layered with dust and dirt. Lab tests revealed the presence of insect eggs (bed bugs), fungi, and bacteria, mostly of low virulence, with the significant exception of the fecal spore-forming *Clostridium perfringens,* which can cause gangrene.

Reconditioned mattresses make up ten percent of the U.S. mattress market. Because of their lower cost they are used in places such as motels, hotels, and nursing homes, as well as in private residences. Only a few states have regulations governing reconditioned mattresses, and fewer still enforce their regulations. The potential for catching an infectious illness from sleeping on a reconditioned mattress is something I'll return to in a little more detail in Chapter 7. For now, let me note two effective **Protective Response Strategies.** The first would be to invest in new mattresses, money well spent if household finances allow. The second is to seal all mattresses inside plastic or heavily woven, impervious coverings, which will create a barrier against infectious germs and debris.

New mattresses or reconditioned ones, the alarms go off and the members of the household start their day. From a contact-spread perspective, the next question arises as soon as their feet hit the floor. Is the floor covered by wall-to-wall carpeting or an area rug, or is it a bare hard surface? All carpeting and rugs retain a variety of germs and debris, including sloughed-off skin cells, pet hair, fungi, mites, and other allergens. Foot traffic from outside and normal indoor activity continually add new germs and other material to this mix. Infectious germs can thus easily find food and shelter in the fibers of a carpet or rug. Much the same logic applies to heavy draperies, textile wall coverings or hangings, and upholstery, which can also become collection areas and breeding grounds for infectious germs, as well as a source of numerous allergens.

Wall-to-wall carpeting is less healthy than area rugs because it covers so much more surface area. It effectively transforms an entire floor surface into one extensive fomite. In addition to covering fewer square feet, area rugs can be cleaned more easily, especially by being taken outside and beaten. In general, the healthiest floor surface for any room in the house is a bare, hard one, because it is the easiest to keep clean.

There are obvious tradeoffs here. Rugs and carpets can add

beauty, warmth, and comfort to a house. But their potential to harbor infectious germs calls for one or more **Protective Response Strategies**. The first of these might be to forgo wall-to-wall carpeting in favor of area rugs. In either case a hygienic household will require the regular use of a good vacuum cleaner with an air filter that will prevent the machine's exhaust from recirculating germs throughout the house. (Likewise, draperies and so on also need regular cleaning.) In Chapter 4 I mentioned the case of a family that repeatedly fell ill with gastrointestinal problems because a leaky vacuum cleaner kept resuspending *Salmonella* germs from the carpets into the household air. Given how common leaky vacuum cleaners are, this sort of thing probably happens far more than is realized. Finally, one could adopt the hygienic Japanese practice of having separate footwear for outdoors and indoors, and leaving the outdoor shoes at the threshold. The effectiveness of this strategy will of course be limited by the extent to which visitors continue to track in dirt and germs from outside. The connection between cultural practices and health is a very strong one, and it will resurface quite a bit in what follows. Some Western practices are healthier than those of non-Western cultures, and vice versa. This is an important issue for public health in the United States because of its increasingly diverse population, as we will see.

Whether on an area rug, a bare floor, or carpeting, the first steps of the day usually take a person to the bathroom and even more specifically to the toilet. When I spoke about proper toilet technique in Chapter 4, I left out one key **Protective Response Strategy**: Always close the lid of the seat before flushing. The reason is that flushing the toilet can send small drops of aerosolized fecal matter as far as twenty feet into the air, potentially hitting every surface of even a large bathroom. Studies have found that flushing the toilet with the lid up can hurl droplets of aerosolized matter onto toothbrushes, combs, and brushes, as well as faucets, sinks, and counters. Obviously, this problem will be all the greater if the toilet bowl is dirty; both the bowl and the seat should be sanitized weekly with a good germicide such as a solution of bleach and water. It also follows that if you're using a public rest room where the toilet seats don't have lids, you should be prepared to exit the stall immediately after flushing.

Now it's into the shower or bathtub, where another germ question awaits. How long has that loofah or bath sponge been in use? We shed about 1.5 million skin flakes every hour, so it doesn't take long for a loofah or sponge to become laden with them and thus with such germs in the normal skin flora as *Staphylococcus aureus,* which can cause infection if it gets into the wrong part of the body. In an environment as humid as the bathroom, a single bacterial cell can multiply into one billion clones overnight. Some germs will even breed on wet bars of soap! To counter these hazards, switch to liquid soap dispensers and thoroughly dry loofahs and sponges between uses. Remember, most germs prefer humid conditions, and drying helps to kill them off or prevent their growth. Loofahs and bath sponges should also be cleaned weekly with a solution of bleach and water.

For the same reason, toothbrushes should be rinsed clean and air-dried in a rack, never placed wet into a case. Razors should also be carefully rinsed and dried after each use, because the blades can capture cells and debris that might provide food for growing germs. And don't expect the rim of the mouthwash bottle to be sanitary. Although the contents are antiseptic, people drinking from the bottle, a very common practice, will have left a plethora of germs on the spout. It is always best to use a disposable paper cup instead.

One last item in the bathroom needs to be considered. It's that nice fluffy towel, fresh from the wash. It may smell and look clean, but if it was washed with underwear, it can easily have become contaminated with fecal matter. The permanent-press and cold-water temperatures that many people use to save energy and preserve delicate garments will not eradicate germs. Combining different kinds of garments in one load to save time only exacerbates the problem. Another factor in the equation is that people use less bleach than they used to because of the popularity of one-step detergents with brightening agents.

About a decade ago I did a study on germ survival in washing and drying machines and found that many bacteria emerged from them unscathed. At a recent meeting of the American Society of Microbiology, researchers reported on a similar study, which found that viral as well as bacterial germs can survive the rough and tumble of being washed in warm or cold water. Even on high heat, a dryer does not normally create a hot enough environment to kill the germs that may be clinging to garments. These germs include fecal bacteria such as *E. coli* and disease-causing viruses such as rotavirus, adenovirus, and hepatitis A, which also have a fecal origin. In fact, a person can theoretically become infected by moving wet clothes from the washer to the dryer and then touching the nose, mouth, or eyes, or preparing or consuming food or drink, without washing hands first.

To keep "clean" laundry from making you sick, follow this **Protective Response Strategy:** Wash underwear separately in hot water (at least 155 degrees Fahrenheit). Add bleach to all wash loads containing underwear, or use a detergent containing one of the new sanitizing agents. Periodically run an empty cycle with bleach and water in order to disinfect the washing machine itself. These steps will insure that your laundry really comes out of the wash as clean as you want it to be. Finally, if climate and other circumstances allow, consider line-drying wet clothes in the sun rather than putting them in the dryer. The sun's UV rays do a wonderful job of killing germs that may be lingering on the laundry. Have you ever smelled clothes after they've dried on the line in the sun? If you have, I'm sure you'll agree that that clean, fresh smell is unforgettable.

After that detour into the laundry room, let's rejoin our typical household on a typical morning. By this time the family members have begun to gather in the kitchen, which can provide a fertile breeding ground for a vast array of germs. Many of the health issues in the kitchen surround handling contaminated food, a type of common-vehicle transmission that I'll discuss in detail in the next chapter. For now, consider all the kitchen items, from aprons and dish towels to sponges and can openers, that people constantly pick up and handle, depositing germs that can easily be transmitted to the next person who touches them. Then there are the tabletop, countertop, chairs, and so on, all of which can be spattered with any cold, flu, and stomach viruses that the house's inhabitants are harboring, as well as with normal respiratory and skin flora. Shocking though it may seem, the kitchen is generally dirtier than the toilet. A toilet is always being flushed, after all.

Even if it contains fecal matter, that fecal matter may not contain any pathogenic germs. But a kitchen almost inescapably does contain pathogenic germs, because of the presence of pathogens in raw meat and other raw foods. For instance, poultry may contain *Salmonella* or *Campylobacter,* and beef may harbor *E. coli* 0157.

Regular cleaning of floors and surfaces, proper food and utensil handling, and frequent hand washing are the main means of lowering risks of infection in the kitchen. It's a good **Protective Response Strategy** to get into the habit of washing your hands after every stage of preparing a meal or dish. It's also important to wash utensils and cutting boards between tasks. For example, never use a knife to cut up vegetables after you've used it to cut meat, unless you've washed it first. Rags and sponges should be washed with soap and water, or a solution of bleach and water, rinsed clean, and then left to dry. Alternatively they can be put in a dishwasher or, when moist, in a microwave for one to two minutes. Otherwise they may hold food residues that will serve up a veritable banquet for germs.

These measures should keep the kitchen clean, unless the cook or someone else who uses it regularly is constantly spreading disease-causing germs like the famous Typhoid Mary. An Irish immigrant whose proper name was Mary Mallon, Typhoid Mary worked as a cook in various New York City households under assumed names in the first years of the twentieth century. She had to use false names, because she made people sick wherever she worked. Although she herself had no symptoms and remained healthy, she carried the germ that causes typhoid fever, *Salmonella typhosa (typhi),* and she transmitted it to her employers, their families, and their guests in the puddings, cakes, and other dishes she made and through the things she touched. New York City health authorities eventually detained her in a hut on the grounds of an isolation hospital on North Brother Island in the East River. She sued for her release, which was granted after a few years on the condition that she work not as a cook but in a laundry, where she was to handle only the dirty clothes. But the kitchen was her calling, and in 1915 she went back to work as a cook under a false name in, of all places, the kitchen of a maternity hospital. A fresh outbreak of typhoid fever followed, including two deaths. She was apprehended again and again quarantined on North Brother Island, with only her dog for company, until her death twenty-three years later. In all, Typhoid Mary infected more than fifty people with typhoid fever bacteria and was responsible for three deaths. So if you're scrupulous about cleaning and people are still getting sick, make sure everyone in the house gets checked out by a doctor.

In the rest of the house there are carpets, rugs, drapes, and upholstery, the germ risks of which we've already seen. There are also items that everyone in a house will touch, like the TV remote control and the telephone. Even children who are too young to make phone calls except by accident or as a prank will be attracted to the phone by seeing older family members constantly using it. A telephone handset soon becomes coated with respiratory and skin germs. When we're in couch-potato mode, the remote control readily acquires a veneer of oils and other residues from the food we're eating while watching television, creating a food source for germs. The **Protective Response Strategy** here is to clean items like the telephone and remote control weekly, to remove the bioloads that steadily accumulate on them. Anyone with a cold or other infectious illness should refrain from using the phone unless it is absolutely necessary, and in those instances the phone should be sanitized immediately afterward with an alcohol wipe or antiseptic spray.

If there are children in the house their toys are likely to be scattered all about and to harbor their own germs and those of their playmates. It would be a perfectly usual thing, for example, for the seat and steering wheel of a toddler's go-cart to be coated with germs ranging from normal skin flora to beta-hemolytic streptococcus group A, which can lead to strep throat or flesh-eating disease, and cold, flu, and stomach viruses. Later on in this chapter, we'll look in some detail at the special germ risks in a child's environment and how to deal with them.

Let's assume that thanks to luck or good Protective Response Strategies, the members of this typical household have gone through their morning routines without picking up any infectious germs inside the house. As they walk outside and close the front door behind them before setting off for work, school, errands, or play, it may seem that they have now left the house's germ risks behind. Not quite. The risk we haven't yet considered is right at the front door on its handle, which may well bear a lingering infectious germ deposited by a member of the household or a visitor. Similarly, the doorbell or knocker may present a risk, however small (time and the elements take care of most of this risk), to visitors or to children who are not yet old enough to have their own house keys. This underlies the need for having a regular schedule for cleaning the entire house, or at least all the normally touchable surfaces of it.

Before concluding our germ tour of a typical house and moving on to the different risks of contact transmission that adults and children face outside the home, I want to mention one more aspect of communal living. Anyone who lives with other people knows that if one person gets sick with an infectious illness, the others are soon likely to do so. This is especially true in families with young children, who catch infections more easily because of their still immature, developing immune systems. Unless family members are going to be so stiff and cold as to avoid touching each other, some of this is inevitable. (I hasten to add that a health-conscious family can dramatically reduce the incidences of shared infections by following the Protective Response Strategies in this book, and still express warm physical affection for each other. That kind of contact is vital for our emotional and physical health.) But it raises the question of how much physical intimacy with others is normally desirable or appropriate, especially in a society like ours, where handshakes between total strangers are common business etiquette and casual friends often greet each other with embraces and face kisses. The polite bow of some Asian and Arabic societies is a much more hygienic practice than our handshaking and face kissing. We'll need to keep the cultural dimension in mind, along with appropriate response strategies, as we consider the potential for contact transmission in the public environments, from offices and classrooms to buses, taxis, and commuter trains, that people frequent everyday.

Is It Safe to Go to Work?

The modern office is densely populated with fomites, objects that can harbor infectious germs and that many different people touch in turn. For a segment on ABC's "20/20" I once took germ cultures from numerous sites within a large office complex in the Wall Street area. These sites included phones, tables, desktops, lights, computers and keyboards, copying and fax machines, door handles, and food vending machines, among others. The results showed the same rich profusion of environmental and human germs that we saw in our tour of New York as germ city and our survey of a typical suburban house.

If we try to count up the number of potential germ transfers that are possible in the average office environment, we'll soon reach a staggering figure. Just think of the number of times in the course of a single day that someone in an office—a colleague, a temporary worker, or a visitor—will use the phone on someone else's desk and grab a pencil to jot down a note or message. That person can't move on without leaving germ mementos behind on the phone, the pencil, and anything else he or she has breathed on or touched. If these mementos include cold, flu, or other infectious germs, those who touch the same objects afterward can easily extend the chain of infection by then touching their eyes, noses, or mouths.

In this regard offices may well be becoming less healthy. In the modern office private work spaces are disappearing even for upper-level managers. There is an increasing trend toward open-plan interior layouts where desks, cubicles, and conference rooms are assigned on a daily basis and lockers are provided for personal belongings. This business-as-high-school setup is being adopted not only in cutting-edge technology firms, hip advertising agencies, and so on, but also in traditional, well-established companies. If you work in an environment like this, you really have to wonder how thoroughly the office is cleaned every night, because you have no way of knowing who was using the computer, phone, and so on before you.

Although they may not understand the science of how infections spread in the work environment, office managers and human-resources executives see the results of it every week or month in the toll of employee sick days. Eighty percent of the acute illnesses that occur in the United States every year are respiratory infections caused by common cold and flu viruses. Bacterial

infections can be treated with antibiotics, but for the most part
viral infections simply have to run their course. The population as
a whole contracts two to four such infections per person per year.
That number includes some ninety-three million common colds,
most of which are caused by rhinoviruses and coronoviruses. The
main symptom is an inflammation of the mucous membranes of
the nose ("rhinovirus" literally means "nose virus"), but that is
more than enough to make a cold victim miserable and account for
the annual loss of 300 million person days of work and school.
Sometimes colds set the stage for complicating sinusitis infections,
which devour another million work and school days annually.

For their part flus cost Americans $10 billion a year in lost
wages and medical expenditures. There are actually three types of
flu virus, influenza A, B, and C, and constantly changing strains
within each type. The influenza A virus, which can infect animals
like birds, pigs, and horses as well as human beings, has been
responsible for great loss of life. In the so-called Spanish flu pandemic of 1918–19 (a pandemic is a very widespread epidemic),
more than twenty million people died. Although better medical
care and understanding of how flu bugs migrate from person to person has lessened the likelihood of such killer pandemics, vicious
type-A flu viruses seem to pop up cyclically, about every ten years
or so. The other flu viruses, B and C, infect only humans in nature
and spread mostly during the winter months, infecting about five
percent of the population. Every year from 30 to 120 million people
worldwide get the flu, depending on the virulence of the virus
strains involved. In the United States alone, 20,000 to 50,000 people die every year from the complications of influenza infections.
Ninety percent of these deaths occur when flu paves the way for
pneumonia in persons sixty-five years of age or older. This is why
a flu shot every year can be a lifesaver for older people.

We all know from personal experience that most respiratory
infections from cold and flu viruses occur during the winter. As
much as one-half of the population has a respiratory infection during the midwinter months, whereas only about ten percent do so
in midsummer. The exact reason for this is uncertain, but the
prime cause seems to be that indoor heating dries out our mucous
membranes and makes them more vulnerable to infection. A likely

contributing factor is inadequate ventilation when windows and doors have been shut tight to keep out the cold. The average adult inhales fourteen pints of air, and about eight germs, every minute (that's ten thousand germs a day). When you inhale germs, the mucociliary linings of the respiratory tract entrap or sweep away most of them. Any germs that make their way into the lower lungs are usually engulfed, or phagocytized, by special blood cells called macrophages. The body repeatedly puts out brushfires in this way, luckily for us. But the body's natural defenses may not be enough in a wintertime home, office, or school, where poor ventilation inevitably raises the concentration of viral respiratory secretions in the air and on fomites.

Stress also plays a role in colds and flus, and for most people summer is a less stressful time than winter, when schools and businesses are running full tilt. The shorter, darker days of winter also add to the stress of the large numbers of people who suffer from seasonal affective disorder or depression, known as SAD. In a recent study, fifty-five adult volunteers were infected with a cold or flu virus, after being interviewed beforehand to assess their psychological stress levels. Researchers found that volunteers who reported greater psychological stress experienced greater physical stress when the cold or flu hit: more severe respiratory symptoms, greater mucus production, and higher levels of interleukin-6, a protein that helps coordinate immune responses. This clearly implies that stress can intensify cold and flu symptoms.

On the other hand, some personality types seem to be less susceptible than others to getting sick in the first place. Another study found that moderately aggressive or assertive individuals seem to have faster-acting, stronger immune systems than meek people. This may have an evolutionary cause. An aggressive person is likelier to be injured by the consequences of his or her behavior than an unaggressive person is. Thus natural selection would have improved the chances of survival and reproduction for those aggressive people with stronger immune systems, building a link between the two traits as one generation succeeded another.

These are all factors you might want to consider next time you feel sick and wonder whether or not you should go in to work. Many of us, unfortunately including some unenlightened bosses, think we're being wimps if we stay home with a cold or flu. Afraid of being labeled as malingerers or poor team players, or unable to accept the thought that we're not indispensable for two or three days, we heroically go in to work when our bodies are telling us to stay home and rest. Once we've arrived in the office, we struggle through the day, accomplishing little except to give the germs that are making us sick the chance to find five, ten, or fifteen more victims.

But heroism is supposed to be about self-sacrifice, not sacrificing others. Suppose you go in to work with the flu and infect a soon-to-be-retiring coworker in her sixties, whose immune system is not what it was in her youth. Why take the risk that your flu germs will start a disease process in this person that progresses to pneumonia and perhaps even death? When we're sick, we need to sacrifice our egos, put macho expectations aside, and stay home to rest and get well. That will create a win-win situation for everybody: We'll recover faster, and the work of the office will be more effective because fewer people will be out sick. In short, real heroes stay home in bed when they're sick.

The appropriate **Protective Response Strategy** for colds and flus is multipronged. It applies not just to the workplace, but also to most other life situations, public and private. Boiled down, it really amounts to nothing more or less than the golden rule: Do unto others as you would have them do unto you.

- **When you have an infectious respiratory ailment, stay home and avoid close contact with others until you are no longer contagious, which usually means until after acute symptoms and fever subside. If circumstances require that you absolutely must leave the house, or if you live with others, then carefully practice the following behaviors:**

- **Wash hands frequently, especially after contaminating them with a cough or a sneeze, or after using a bathroom facility.**

- **When you sneeze or cough, cover your mouth and nose with a tissue, which should immediately be discarded. Handkerchiefs are fine for mopping up perspiration, but** they become germ reservoirs when we use them to blow our noses and then stuff them back into a pocket or purse.

- If you have to sneeze or cough and you do not have a tissue, sneeze into the crook of your arm or double cup your hands over your mouth and nose. Then wash your hands as soon as possible.

- Avoid crowded public areas as much as you can.

- Buses, trains, and taxis abound with surfaces, such as handrails, that act as collecting areas and transfer points for germs. After touching these surfaces, try not to touch your mouth, nose, or eyes until you have had a chance to wash your hands.

- Use a tissue or paper towel to open and close doors, including bathroom doors, so that you do not contaminate yourself with other people's infectious germs, or contaminate the door handles with your own.

- For influenza A and B infections, consult with your doctor about taking one of the oral and nasal vaccines which will become available in the near future. They may prevent or moderate symptoms.

This **Protective Response Strategy** has clear implications for those who are not sick and who want to protect themselves from catching someone else's cold or flu. If you are working alongside sick people, your protective options are limited. But you can help yourself stay healthy by being alert to contamination from fomites and by washing your own hands frequently throughout the day. As I've discussed in earlier chapters, frequent hand washing is a good daily habit no matter what circumstances surround you.

In this connection, it is important to remember all the handshaking, embracing, and face kissing that go on in our society. Whenever possible, handshaking and face kissing should be reserved for close friends and family, so long as they show no evidence of being ill. Face kissing will not likely transmit germs under normal conditions, as mouth-to-mouth kissing will, but it represents a risk whenever one of the people involved is ill. And it isn't always easy to know when this is the case. People harboring flu viruses, which spread more easily than cold viruses, can pass them to others one to three days before, and up to five days after, they themselves develop any symptoms. The upshot is that after you've just shaken hands with a new business acquaintance and sat down for a meeting, you'll want to avoid touching your eyes, nose, or mouth until you've washed your hands.

The communal impact of a single person's illness suggests two additions to the cold and flu **Protective Response Strategy**. They are as important as they are hard to implement.

- If you see that a close coworker has come in sick, kindly encourage that person to go home and rest for the sake of all concerned. If the person insists on staying on the job, politely point out any disease-spreading behaviors they may engage in, such as failing to cover a cough or a sneeze with a tissue, using someone else's phone, and so on. Increasing numbers of people, especially young people, just don't realize that these behaviors endanger others. If you can't do this without lecturing or scolding, or if it's just not advisable politically, then of course you may have to keep silent—and as much as possible keep your distance.

- If you work in an organization that does not appreciate the realities of infection, try to educate others about them. Allowing for circumstances (the world being what it is, you can't always speak your mind plainly), diplomatically let people know that, at least in regard to infectious illness, macho attitudes about "toughing it out" and "taking one for the team" only make the team less efficient and productive. Smart companies will recognize these realities and make them a part of all employee training and protocols, saving untold millions of dollars in medical costs and lost workdays.

The interaction of germs and people always has a public and a private dimension. Our personal efforts cannot protect us if too many of those around us are not also protected. Effective solutions to infectious illness thus require both community and individual action. The bottom line is that our health, whether on the level of the individual, the family, or society as a whole, depends on our sense of responsibility and concern for each other. We can't save the whole world at once, of course, but we can improve the world's health one person, and one person-to-person interaction, at a time.

Dirty Money

There is virtually no limit to the number of objects that can become infection-bearing fomites. But one that deserves special mention is money. Thanks to the illegal drug trade, the currency in the average person's wallet or purse bears trace elements of cocaine. Thanks to the mechanics of contact spread, that same currency is also permeated with microbes. In our fast-food society, people regularly handle money and then eat or drink without washing hands and without thinking about the risk of infection. But the risk is real.

Consider this imaginary, but very possible, scenario. One summer morning a hot-dog vendor goes to work with a cold (colds happen in summer, too, after all). He begins work by retrieving his cart, which has sat idly overnight, ample time for germs to flourish on food particles or drippings from the day before. As the vendor prepares his wagon for the day, he turns up the heat and then places hot dogs into hot-water tanks. Then he coughs into his hands, and blows his dripping nose into a tissue. Without washing his hands, he arranges packages of buns and fills up tubs of relish, mustard, sauerkraut, onions, and other toppings, transferring his cold virus to all of them. Next he loads four cases of canned soft drinks into an ice chest; as he takes cans out to sell to customers, they will drip water onto the food cart, providing moisture for growing germs. Finally the vendor counts his start-up money with wet hands, enhancing germ growth on the bills. Then he heads to his station on a busy city street.

One of a street vendor's biggest challenges is the need to visit a rest room every few hours. Groups of vendors employ a floating substitute, who spends his day spelling each vendor in turn. In other cases a vendor may relieve himself in a container in a car or van that he has parked nearby. Under both sets of circumstances, personal hygiene is poor at best.

Over the course of a humid summer workday, germs will have many opportunities to grow on the vendor's cart. As the vendor sells food and drink to his customers, spattered condiments, melting ice water, and meat drippings provide food and drink for germs. These germs will inevitably find their way into both customers' food and into the fibers of the paper money that is exchanged at every transaction. The germs will remain on the money as it travels through the rest of the daily economy in shops, newsstands, and so on.

Many studies, including two of my own, have shown that money can be an effective fomite for germ transmission. ABC's "20/20" asked me to help them prepare a segment on this issue, and I devised a plan for collecting money from street vendors, shops, restaurants, and other establishments in Chicago, New York City, and Washington, D.C. After each transaction, the bills received in change were put directly into newly purchased wallets, which were then sealed in plastic. The bills were tested and found to be contaminated with germs of fecal, respiratory, and skin origin. Although the risk of contracting a serious infection from dirty money is low, the germ count is high enough to make it easy to contract a cold, a bout of diarrhea, and similar ailments.

The **Protective Response Strategy** needed here is, once again, frequent and thorough hand washing, especially before eating and drinking anything. In a pinch, if you can't get to a sink to wash with soap and water, a moist towelette or a waterless liquid germicide, either of which can be bought in any drug store, can be used instead. But there is no substitute for proper hand washing.

The Common Pool

Although most of us find that a sense of community is sorely lacking in modern life, that does not mean that there is any shortage of common contact, direct or indirect, with other people's germs. Spas, saunas, and showers in health clubs should all be posted, "User beware; people deposit germs here daily."

Beauty salons and barber shops can be particularly dangerous areas for germ contact, because the equipment in them often gets used many times in a row without being sanitized between customers. For example, investigators from the CDC recently discovered that pedicure footbaths in nail salons frequently teem with *Mycobacterium fortuitum,* a relative of the germ that causes tuberculosis. *Mycobacterium fortuitum* can cause boils that resist treatment with all but a few antibiotics. Some women who were infected with *M. fortuitum* because of a pedicure footbath in a nail salon suffered scarring that had to be treated surgically with skin grafts. In California, where this phenomenon was first identified, officials are drafting regulations that will require nail salons to clean and disinfect footbaths regularly. In one California nail salon, a single pedicure footbath infected at least 110 customers with *M. fortuitum.* But in the meantime, and for those who don't live in California, caution and common sense should be your guide. Don't be afraid to ask when a footbath, a razor, a whirlpool tub, or the like has last been cleaned. Your health depends on it.

Children and Germs

Because their immune systems are immature, childen under six are at increased risk of infection from contact spread, as well as other forms of germ transmission. As I discussed in Chapter 4, a baby acquires the first germs in its normal flora from its mother's birth canal, skin, and respiration. This transmission of microbes from mother to child is essential to the baby's health. In fact, routine exposure to environmental germs helps to develop a normal immunologic system. Likewise, low-birth-weight babies develop fewer infections when they are breast-fed rather than bottle-fed, because this allows the mothers to pass on valuable antibodies. But if the mother harbors potential pathogens such as group B streptococcus, *Listeria monocytogenes,* or *Hemophilus influenzae,* the baby could be harmed by exposure to them during birth or after.

As children grow into the toddler stage, their immune systems get additional boosts if they receive vaccinations against childhood illnesses. Nevertheless, children remain subject to numerous infections. One of the most common is otitis media, an infection of the middle ear, which leads to thirty million doctors' visits every year. The cause of the infection is often an antibiotic-resistant strain of *Streptococcus pneumoniae*. Children easily pass these potentially disease-causing germs, and others they frequently harbor, such as *Hemophilus influenzae* and groups A and B strep, to other children and adults.

Sixty percent of the nation's children are under six years old. That translates to about thirteen million children, many of whom spend time in day care and preschool. Many studies have shown that children in such settings are at significantly greater risk for diarrheal disease. Diaper changing is probably the greatest single means of contact transmission of a child's germs to caregivers, teachers, and other children, as well as family members and others in the local community.

Many viral agents can lead to diarrheal illness, which may include vomiting and dehydration. One of the most problematic in young children is rotavirus. Outbreaks of rotavirus in a day-care program or preschool sometimes require that the school be closed and even that children be hospitalized. Worldwide the virus, which also affects immature animals such as calves and piglets, is a major cause of death in children under five years of age. It is estimated that in the underdeveloped world children under five suffer three to five billion cases of diarrheal illness annually. Although the disease is self-limiting and usually lasts one week or less, up to five million children under the age of five die from it worldwide every year.

A segment on "Oprah" provided me with an interesting opportunity to survey the germs in the environment of some average American children. I joined a group of mothers and their children, aged one to fourteen, in a suburban home on Long Island. The children's mothers and I then collected thirty-four germ cultures from such places as the seat of a child's ride, a shopping-cart handle, toys, an infant's walker, a vending machine, a movie-theater seat, a swing set, a water fountain, a desk, the shoe lockers at a fast-food restaurant's play area, a kitchen table, an ice cream parlor countertop, school lockers and gym mats, and the shoes and balls at a bowling alley.

Believe it or not, if you don't already suspect it from our survey of "Home Truths," but the heaviest microbial growth was on the kitchen table in the house where we met. Next heaviest were the steering wheel of a child's ride, the shopping-cart handle, the vending machine's coin slot, and the infant's walker. These objects bore numerous fecal and skin germs such as *E. coli* and *S. aureus,* respectively. In addition, the steering wheel was covered in beta-hemolytic group A strep, which can cause strep throat or flesh-eating disease; the infant's walker had a heavy growth of *S. aureus,* as did the bowling shoes, where the germ was creating a foul stench; and the movie-theater seat and the thumbhole of a bowling ball contained substantial amounts of fecal contamination.

The presence of so many germs in a child's environment calls for two **Protective Response Strategies**. The first is regular cleaning of the toys, objects, and surfaces that young children regularly touch at home. For most situations, it is best to use innocuous, nontoxic cleansers such as hydrogen peroxide or soap and water. But if you suspect that an area is highly contaminated (a bad smell is often a good indicator of this), you can use a mild bleach solution (one part bleach to nine parts water) or a commercially available germicide; be sure to apply the germicide for the recommended period of time and to rinse the item thoroughly.

The second **Protective Response Strategy** is to train children in good hygiene, so that they can learn to protect themselves not only at home, but at their friend's homes, school, the mall, and elsewhere. Children should be taught to:

- **Wash hands before eating or drinking, and after using a bathroom facility.**

- **Never use someone else's eating or drinking utensils, even if the other person is the child's best friend.**

- **Wash dirty toys and objects that have touched the ground.**

- **Never eat food that has fallen on the floor or ground.**

- **Wear long pants or skirts when visiting public facilities such as a movie theater, where sitting down with bare legs could lead to contracting an infectious illness from germs on the seat.**

- **Practice proper cold and flu etiquette, such as covering one's mouth and nose during a sneeze or cough. (The Protective Response Strategy for the workplace applies in full force to schools, where countless fomites are contaminated by children, teachers, and staff. Sick children should stay home from school until they're fully recovered.)**

All children need to learn these lessons. But not every family in our increasingly diverse society is able to teach them. Because of our current high levels of immigration, more and more children from poor, Third World backgrounds arrive in American schools without knowing basics such as how to brush their teeth, wash with soap and water, or shampoo their hair. Some have never even seen a toothbrush until a teacher demonstrates how to use one. Because of what the *New York Times* calls "a flood of immigrants to classrooms," school systems across the country face the challenge of instituting special programs to address what is becoming a burgeoning public-health problem.

Poor hygiene is by no means only a problem of immigration; I noted in Chapter 2 that it encompasses every segment of our society, and I have more to say about it below. But I cannot help making a few observations as both the proud descendant of Italian-American immigrants and as a medical scientist. Largely at the behest of big business, which is always hungry for cheaper workers, Congress has since 1970 steadily increased immigration to the highest levels in our nation's history, about a million people a year, most of them poor. But Congress has not provided for the social services that an expanding population of poor people requires. Many such programs have actually been frozen or cut back as part of an anti-welfare agenda. With rates of immigration four times the historical average driving the population growth of America's poor to Third World levels, there is a growing mismatch between needs and services. This mismatch must be resolved humanely, one way or another. The gap between rich and poor is greater in America, and expanding faster, than in any other developed country, and that inevitably means an increase in the rates of infectious diseases. It is time to put politics and greed aside, and recognize that this is not a healthy situation for anybody.

In terms of infectious illness, we must of course recognize that poor hygiene is increasingly widespread among all young Americans, no matter where they were born. The need for change may be greatest among the immigrant poor, but the scandalous rates of nosocomial infection—the kind that doctors, nurses, and other hospital staff cause by not washing their hands—show that a cavalier attitude to cleanliness cuts across all socioeconomic lines. Children now grow to adulthood thinking that it is okay to cough in someone's face or not to bother washing hands after going to the bathroom.

I can't help feeling that one of the chief culprits in this state of affairs is our society's focus on self-esteem and self-expression. Although these are good things and should certainly be fostered in every child, we seem to have forgotten the importance of learning and practicing self-respect, which is an even more basic human value. It is hard to feel good self-esteem if you don't know how to keep yourself respectably clean and healthy. When I was a child growing up in Brooklyn, my classmates and I, a diverse mix including many immigrants, were regularly drilled in proper hygiene. The country should revisit this issue as part of a national commitment to public health. As I said above, all the protective responses in the world can't keep you safe if too many of your neighbors and fellow citizens are constantly putting themselves and others at risk of infection.

A strong indication that today's children need extra help in coping with infections is that more and more eight-to-ten-year-olds are showing up in doctors' offices and emergency rooms with mononucleosis, the "kissing disease." To some extent this seems to be the result of new strains of mononucleosis germs, which can pass from child to child by touch rather than deep kissing, but earlier experimentation with kissing must also play a role. The incidence of sexually transmitted disease (STD) in general has reached epidemic proportions among teenagers and other young people, many of whom have never been taught the basic infectious facts of life. According to the CDC, about fifteen million people become infected with STDs in America every year, and teenagers account for twenty-five percent of these cases.

Intimate Contact

With numbers like these it doesn't matter what sort of neighborhood you live in. The girl next door and the boy down the block have a sexually transmitted disease. It could be human papillomavirus (HPV), which causes genital warts and strikes about 5.5 million people a year. Or chlamydia, which affects three million people of both sexes annually and can lead to such severe pelvic inflammation in women that ectopic pregnancy and sterility result. Both HPV and chlamydia have been linked to increased risk of cervical cancer.

The rates of herpes 1 and herpes 2 simplex are so great that some of your neighbors, friends, or extended family almost certainly have one of these viruses, which can flare up at any time and produce oral and genital sores. According to the CDC, more than one in five American adolescents and adults are infected. Members of the medical profession sometimes mordantly refer to herpes as "the gift that keeps on giving." The viruses pass from person to person via secretions from the sores. They can remain dormant for an entire lifetime, or some period of stress may trigger them into action. Thus many people who carry one of the viruses never realize that they have it. Although herpes 1 generally affects the upper body and herpes 2 the genitalia, both can be communicated sexually and they often switch places. About twenty to forty percent of new genital infections are caused by HSV-1 and are transmitted by oral sex. The viruses are so closely related that it is usually considered unnecessary to distinguish between them at this time, but herpes 1 recurs less frequently than herpes 2. Planned Parenthood estimates that ninety percent of the American population is exposed to herpes 1 in childhood, but both herpes 1 and 2 can be acquired by any kind of sexual activity, including kissing, if the disease happens to be active in an infected person. Because a newborn infant can contract a herpes virus from its mother's genital tract, pregnant women who test positive for herpes often must undergo cesarean section. If no symptoms are present, however, a woman with herpes can deliver a healthy baby vaginally.

Or maybe one of your neighbors has gonorrhea, which strikes 650,000 people in America every year, or perhaps even AIDS.

Sexually transmitted diseases all have their own unique disease processes, but they do fall into two main groups, bacterial and viral. Medical science has developed antibiotics to cure most bacterial STDs, such as chlamydia, gonorrhea, and syphilis, although antibiotic resistance can make their treatment more difficult. There is no cure for the viral diseases like herpes 1 and 2 or AIDS, however. Some sixty-five million Americans have these incurable viral STDs. While a search for preventive vaccines for these illnesses is under way in laboratories all over the world, doctors and patients must rely on drugs to moderate their symptoms. Some of the newest drug treatments are very promising. For example, the latest AIDS "cocktails," as they are called, have proven effective in patients who have been carrying human immunodeficiency virus (HIV), the virus that causes the disease, for less than six months. These patients can often temporarily stop drug therapy after a relatively short course of treatment; apparently their immune systems have been so well stimulated that they can keep the virus in check. This holds out the hope that medical science will find a reliable way of turning on the immune response as needed, despite the crippling effect that HIV almost always has on the immune system's T cells.

Soon after it was first recognized, in 1981, AIDS became the signature disease of our time. Its appalling death toll has added an element of dread to all our lives, making it the most feared plague in human history. Since the epidemic became full blown in the 1970s (retrospective blood tests have revealed scattered cases decades earlier), AIDS has killed twenty-three million people, as many as died in Europe during the worst visitations of the Black Death, the bubonic plague. Astoundingly, three-quarters of these deaths have occurred in Africa. In the year 2000 alone three million people died from AIDS, eighty percent of them in Africa, making it the leading disease-specific cause of death worldwide. A small but significant number of AIDS cases have resulted from contaminated blood in transfusions, perhaps most tragically in Romania, where hundreds of people were infected with HIV in this way. But the overwhelming majority of cases have two causes: unprotected sex and the sharing of needles in intravenous drug use.

The near future will bring a great many more deaths from AIDS. Worldwide, an estimated thirty-six million people are infected with HIV, and 25.3 million of these are in sub-Saharan Africa. Rates of HIV infection as high as thirty percent are common in sub-Saharan African cities, and the disease is also growing by leaps and bounds in much of Asia, the Caribbean, and Latin America. Infection is spreading fastest among young people. Worldwide, six people below the age of twenty-five become infected with HIV every minute, and girls' rates of infection outpace those of boys in most places, largely because of cultural customs that make girls unable to say no to sexual demands or outright sexual assault. Each year, about six million people are newly infected with HIV. That's about fifteen thousand people a day.

In the context of person-to-person contact transmission of germs, it is fascinating to note that AIDS is an example of what we might call cousin-to-cousin transmission. HIV originated in chimpanzees, the closest species to human beings. In Central Africa chimpanzees are hunted as food, and the virus probably passed to human beings through cuts or other contact with blood during hunting and eating. The virus must have had an easy transition. Out of 3,000,000,000 DNA nucleotides in each species, chimpanzees and humans share 2,999,400,002 in common!

Although the long-term prospects for an AIDS vaccine are extremely promising, as we will see, in the short term we must focus on preventing the disease from occurring. Here the person-to-person angle emerges in an especially complicated way. The incidence of HIV transmission can be dramatically lowered by the simple use of condoms during sexual activity (this also applies to all other STDs) and of clean needles during intravenous drug use. But ignorance and entrenched cultural attitudes present enormous obstacles to public-health campaigns seeking to teach people how to protect themselves and others. In America such efforts often are blocked because of fears that providing clean needles will encourage more drug use; really, it will just save lives, including those of non–drug users who may wittingly or unwittingly come into intimate contact with IV drug users. In some African societies, traditional beliefs about men and women have led to such tragic and horrific behavior as HIV-infected men deliberately having unprotected sex with virginal girls. In the mistaken belief that this will cleanse them of disease, they only spread it further.

As with other infectious illnesses, this calls for a twofold **Protective Response Strategy.** We must do everything we can to educate people, particularly the young, about AIDS and other STDs. We must take public steps to encourage the use of condoms and clean needles. And in our own lives, we must practice safe sex and other safe behaviors, and never have unprotected sex outside the confines of a committed, monogamous relationship.

The connection between disease and culture was driven home for me in the fall of 2000 by news of an outbreak of Ebola fever in the Ugandan district of Gulu. Ebola is another disease that passed to humans from among our close relatives, this time from a certain monkey species. It can be transmitted from person to person by sexual contact. But as we've already seen, this terrifying disease, which literally liquefies a person's insides, causing profuse internal and external bleeding, can also be passed by touch or even respiratory secretions. In Uganda this easy transmission has been enhanced by traditional funeral practices, in which family members and close friends bathe the deceased person and then wash their hands in a communal bowl as a sign of unity. This practice triggered the recent outbreak in Uganda, when a thirty-six-year-old woman, the mother of two children, fell ill and died in her mud hut. Not knowing that she had died of Ebola fever, the woman's mother, her three sisters, one of her children, and three of her friends gathered together to bathe her body as a sign of their respect and love. All of them have since died of Ebola fever. In total this particular outbreak took more than ninety lives, including those of some medical personnel, before it was contained by (Western) common sense procedures for isolating the dead and insuring that caregivers did not contaminate themselves by failing to wear goggles, masks, and gloves. A complicating factor here is that in many underdeveloped countries, basic medical supplies that we take for granted in the West may be in short supply or unavailable. And in the midst of a crisis, overworked doctors and nurses can easily lose sight of protecting themselves as they try to help others.

Pet to Person

The transmission of germs from animals to people is really very common, if not often as lethal as in the cases of AIDS and Ebola fever. Zoonotic diseases, as illnesses that move from animals to people are called, include anthrax, influenza, and West Nile fever. Moreover, this kind of transmission occurs all the time with pets, who become virtually full-fledged members of the family in many households. Although over time pets and their owners can become habituated to and fairly tolerant of one another's normal flora, people still regularly fall ill because of what their pets track in from out-of-doors or, in the case of exotic pets, from remote corners of the world.

With that in mind, here is a **Protective Response Strategy** for pets and the people who love them.

- **When you stroke, handle, or pat your pet, avoid touching your mouth, nose, and eyes until you can wash your hands. Your turtle or lizard, for example, can carry *Salmonella*, a leading cause of gastrointestinal disease.**

- **Make sure your pets' vaccinations are up to date. Dogs and cats can carry rabies and the syphilis-like leptosperosis germ, both of which can easily be transmitted to human beings.**

- **Keep pets well-groomed. A clean, well-brushed dog or cat will suffer fewer fungal infections and be less troubled by fleas and ticks, and thus be less likely to pass the fungi or the germ-bearing pests on to people.**

- Don't let pets stray into the wild. House cats generally are healthier than outdoor cats, for example, because a cat that roams outside and eats birds and mice will be likelier to carry toxoplasmosis, the number-one parasite world-wide. Toxoplasmosis affects two billion people each year. Among other problems, it can cause birth defects in a growing fetus. An indoor cat that eats prepackaged foods will usually not carry toxoplasmosis.

- Wash your hands after cleaning a cat's litter box or using a pooper-scooper for a dog's feces. This will safeguard you against many of the aforementioned problems, as well as *Toxocara,* a roundworm parasite that migrates in the body. At least two thousand cases of this visceral larval migrans disease (VLM) are reported in the United States every year.

- Pregnant women should never clean a cat's litter box or other pet wastes.

- After giving medications to a sick animal, wash up with a germicide. And never let a dog or cat play with you to the point of overexcitement, which invites bites and scratches. In the case of cats, even house cats, a cut that draws blood may increase the risk of infection with *Bartonella,* which causes cat-scratch disease. This organism lives in the fleas of cats, as well as in and on cats and other animals, and can cause swollen lymph gland disease in human beings.

- If one animal in a multi-pet household is sick, wash your hands thoroughly after touching the affected animal so that you don't transfer any infectious germs to your other pets or yourself. It may also be wise to have the vet check all your pets whenever one of them becomes ill. Keep sick pets isolated from other pets.

- Never allow your dog to lick your mouth, nose, or an open wound. Among other germs, they can transmit *Pasturella,* which could prevent a cut from healing or cause some other type of infection.

Does that last bit of advice sound impossible to follow, or impossible to enforce where children are concerned? I have to confess that when my two daughters, Alexandra and Meredith, were growing up, I struggled in vain to convince them to stop smooching with their dogs. To this day I keep reminding them that it is not healthy to let your dog lick your face, especially your mouth area. But what is a father to do? Fortunately the results of dog kissing man, or woman, are usually not very dire. I suppose you could say that when person and (pet)person have bonded for life, let no man try to put them asunder.

7

Common Ground

The Killing Fields

A recent visit to Gettysburg reminded me of the carnage of the American Civil War. It was in many respects the first technological war, fought with new machines like the railroad and the Gatling gun, the first effective machine gun. Every battleground is a killing field, as the military historians say, but in the Civil War the casualties were truly horrific. In one Confederate attack at Gettysburg, the famed Pickett's Charge, only five thousand soldiers survived from a force that numbered fifteen thousand. The battle of Antietam was even worse, not only the bloodiest day in the Civil War but in all of American history. With twenty thousand casualties in all, each side lost more or less twice as many men as all U.S. forces lost on D day in World War II.

Yet the deadliest agents of destruction on the killing fields of the Civil War were not new man-made weapons, but age-old infectious germs that thrived in the man-made conditions of nineteenth-century warfare. The great American poet Walt Whitman, who served as a nurse attending Union soldiers in a hospital in Washington, D.C., observed, "There are twice as many sick as wounded. For every soldier killed in battle about two [die] of disease." The same proportions held sway throughout the Civil War's battlefields, encampments, military prisons, and hospitals. Of the approximately 623,000 Union and Confederate soldiers who died in the war, twice as many perished from infectious illness as from cannon, rifle, and bayonet.

Many fatalities resulted from infections in postsurgical wounds. Although the Civil War (1861–65) spanned the years just after Pasteur began demonstrating the first clear links between germs and disease, which as we've seen would motivate the English physician Joseph Lister's campaign for cleanliness in surgical operations and other procedures, it would be some time before the medical profession in Europe and America cleaned up its act. But what was far worse than doctors with dirty hands and scalpels was dirty food and water, contaminated with germs of fecal origin.

Conditions in Union camps and hospitals were so bad that a Civilian Sanitary Commission was formed to inspect them. Headed by Elisha Harris, a New York physician, and Frederick Law Olmsted, an architect who would later become one of the chief designers of New York City's Central Park, the commission found that campsites were uniformly "filthy, fetid, and poorly drained," creating ideal breeding grounds for "diarrhea, dysentery, and typhoid." Although the commission did not identify germ contamination as the cause of disease—many in the medical profession still thought diseases were caused by miasmas, poisonous vapors from stagnant water and rotting matter—it called for commonsense measures to clean up soldiers' food and water.

The Sanitary Commission's reports mobilized a vast response in Union states. Groups and organizations held "Sanitary Fairs" to raise money, and many women volunteered to become nurses, who until that time were predominantly male. Dorothea Dix, the Superintendent of U.S. Army Nurses, at first demanded that women volunteers be "over thirty and plain in appearance," but as the war dragged on she was forced to accept younger and prettier women, too. Clara Barton, later the founder of the American Red Cross, coordinated donations to the Sanitary Commission, organized an independent nursing corps, and gained fame as the "Angel of the Battlefield." More than three thousand women served as Union nurses. Perhaps the best known after Dix and Barton was Louisa May Alcott, the author of *Little Women*. Her first book, *Hospital Sketches,* drew on her work as a Civil War nurse. Olmsted said of the women nurses, "God knows what we should have done without them, they have worked like heroes night and day."

These efforts to clean up soldiers' camps, prisons, and hospitals, and to provide fresh water and food, lessened but could not halt the toll of infectious diseases. That would have to wait until army supply and medical corps, like other social institutions, assimilated the mid- to late-nineteenth-century breakthroughs of Pasteur, Koch, and their followers. In the meantime, the solider's greatest enemy remained disease, and chiefly diarrheal disease.

Diarrhea is a symptom of several different varieties of food and water poisoning. It is caused by microorganisms that can be passed to others by the fecal-to-oral route through direct or indirect contact of person to person or animal to person, and through common-vehicle contamination of food and water. Among the germs that can cause diarrhea are bacteria such as *Campylobacter, Clostridium perfringens, Salmonella,* and *Shigella;* protozoa such as *Cryptosporidium, Cyclospora,* and *Entamoeba histolytica;* and numerous flu and stomach viruses. As I mentioned in Chapter 5, a strain of the normal human intestinal germ *E. coli,* known as *E. coli* 0157:H7, was at some point virally hijacked by *Shigella* toxin genes and became a potential source of diarrhea. Dysentery, which involves an especially severe, protracted form of diarrhea, is caused by either *Shigellae* bacteria or *Entamoeba histolytica* protozoa. Because it can rapidly lead to dehydration, diarrhea can be especially dangerous for young children, the elderly, or anyone whose immune system is suppressed or compromised. If diarrhea continues unchecked for too long, it can threaten kidney, liver, and pancreatic function.

At one time or another more than ninety percent of the two million men who served in the Union armies were stricken with what they jokingly called "the Tennessee Trots" or "the Virginia Quickstep." The soldiers' humor had a gallows edge. Although most cases of diarrhea, then as now, were self-limiting, the condition could become very serious in soldiers weakened by wounds, other injuries, or poor nutrition. For soldiers sheltered in unsanitary conditions, there might be no escape but death from repeated exposure to diarrhea-causing germs in tainted food and fecally contaminated water. On the Union side alone, diarrhea killed 44,500 men. Statistics on the Confederate side were no more heartening.

Civil War–like conditions of filthy water and food and inadequate sanitation no longer exist in most of the developed world. But they still prevail in much of the undeveloped world. As I noted in Chapter 2, almost two and a half billion people, forty percent of the world's population, do not have clean water and decent sanitation. The cost of that in lives runs as high as five million deaths annually, most of them young children. The cost to fix the problem would, again, be about $10 billion a year, or about half of what Americans spend annually on pet food.

Even in highly developed industrial countries, diarrhea remains a serious public health problem. Every year common-vehicle food- and water-borne illnesses give diarrhea to approximately 100 million Americans. The cost of this unfortunately must be measured in both money and human life: about $8 billion and at least ten thousand deaths annually. What makes these numbers shocking is that proper hand washing and food hygiene would cut them in half. Greater vigilance by regulatory agencies and the food industry could reduce them even more.

Fast Food, Fast Germs

Examples of what can go wrong with our food and water regularly make front-page news. A typical case played out in Chicago in 1993, when a local Jack in the Box restaurant served undercooked hamburgers to customers. Within a short period of time, hundreds of people fell ill with bloody diarrhea and four children died. The culprit was *E. coli* 0157:H7. This *Shigella*-toxin-bearing strain of the normal intestinal germ *E. coli* first came to light after it was found to be the cause of two epidemics of food poisoning at McDonald's franchises, one in White City, Oregon, in December 1981 and the other in Traverse City, Michigan, in May 1982.

The *Shigella* toxin is only slightly less dangerous than the botulism toxin, which is one of the deadliest known. *Shigella* toxin can damage blood vessels and kidneys with sometimes fatal results, especially in children, the elderly, the immunosuppressed, and the immunocompromised. A recent study suggests that children who have contracted *E. coli* 0157:H7 should not be treated with antibiotics. It appears that antibiotic therapy actually increases the risk that an *E. coli* 0157:H7 infection will threaten the kidneys with what is known as hemolytic uremic syndrome (HUS).

Food-related *E. coli* 0157:H7 infections may be on the rise. It used to be thought that such infections struck about twenty thousand Americans annually, killing about five hundred of them. In December 1998 the CDC revised that estimate upward, doubling it to forty thousand cases annually. It is unclear whether this revision was overdue or made necessary by a sudden increase in cases. Up to fifty percent of cows normally have *E. coli* in their guts. A recent U.S. Department of Agriculture report found that nearly half the beef cattle in American feed lots carry it. Whenever cleanliness and butchering techniques are less than optimal, a cow's meat can easily be contaminated by its own intestinal contents during slaughtering. If that meat is then cooked improperly, it can pass diarrhea-causing germs on to anyone who eats it.

Apparently, you don't have to eat contaminated meat to get diarrhea from a cow, however. You can likely also get it from fruits and vegetables that have been fertilized with cow manure. Common fruit flies can act as a vector for *E. coli* 0157:H7, transmitting it from one piece of fruit to another. One or both of these forms of transmission probably explain recent findings of *E. coli* 0157:H7 on alfalfa sprouts and lettuce and in unpasteurized apple juice. In addition, it is possible to contract the germ from touching animals at a petting zoo (or touching each other, for that matter) and from ingesting fecally contaminated water in swimming pools or municipal water supplies. Public drinking water supplies can be tainted when storms and floods cause sewers to overflow and spill into lakes and reservoirs.

Later in this chapter I will outline a comprehensive Protective Response Strategy for *E. coli* and all other food- and water-related germ threats. First let's see what some of the other threats are.

Germs on Tap

When the hamburgers at a fast-food restaurant are tainted by infectious germs, hundreds of people can easily be made sick before the problem is identified. But if a municipal water supply is contaminated, an entire city can be laid low. That happened in Milwaukee, Wisconsin, in 1993 because of *Cryptosporidium,* a parasitic protozoan germ commonly found in surface water. It usually gets there through the feces of birds, including geese and ducks, reptiles, and many other animals, including human beings.

Over the last twenty years cryptosporidial infections have been associated with intestinal inflammation and diarrhea in such animals as turkeys, rattlesnakes, the European common fox, chickens, guinea pigs, sheep, rhesus monkeys, dogs, cats, horses, pigs, deer, goats, gray squirrels, and raccoons. The first known case of cryptosporidiosis in human beings occurred in 1976 in a three-year-old boy who suffered from severe gastroenteritis for two weeks.

Cryptosporidium potentially threatens all public water supplies. But it is usually present in low concentrations that will not seriously harm most individuals with healthy immune systems. As always with infectious germs, those most at risk are young children, the elderly, the immunosuppressed, such as pregnant women, and the immunocompromised, such as people with AIDS. Only a few cryptosporidia may be necessary to cause an infection in members of these high-risk groups. For example, up to twenty-four percent of AIDS patients will develop cryptosporidial infections, whereas only about five percent of the immunocompetent population will do so. The danger is greatest of all in some areas of the Third World, where there is evidence of *Cryptosporidium* in over ninety percent of the population. It is infection rates like this—remember, *Cryptosporidium* is only one of many germs that can cause diarrheal disease—that lead to some five million diarrhea deaths a year in the Third World. *Cryptosporidium* has now been reported in people from three to ninety-five years old on every continent except Antarctica.

Boiling water for one minute will kill any cryptosporidia it contains. But these hardy protozoa can withstand most chemical attacks. Some cryptosporidia can even survive the hyperchlorination process that many American municipalities use to purify drinking water. If the concentration of cryptosporidia in drinking water becomes high enough, the effect can be almost as devastating in the developed world as in the undeveloped world. In 1993 in Milwaukee some four hundred thousand people, over half the population, became ill with profuse watery diarrhea and abdominal pain; one hundred people died. Since then numerous additional episodes of cryptosporidiosis in the United States have been traced to drinking supposedly safe, treated water from the tap. The scope of the problem has become so great that in 1998 the federal government announced an $800 million water treatment plan to safeguard the nation's drinking water.

Over time this public-works project will bring better filtering, ozonation, and other purifying techniques to reservoirs all across the country. Federal clean-water regulations actually already require drinking water to be filtered. But New York City and several other cities have historically been exempt from this requirement, so long as they could prove that their monitoring systems were sensitive enough to detect potential pathogens. Nevertheless, New York City is planning a $1 billion program to prevent pollution of its most important reservoirs.

Although *Cryptosporidium* is primarily a waterborne germ, it can also be carried by mollusks such as oysters, mussels, and clams. Freezing shellfish for twenty-four hours prior to consumption will kill any cryptosporidia oocysts they contain. But cooking shellfish is really the way to go. Properly refrigerated shellfish can easily contain bacterial or viral pathogens that survive freezing.

The Case of the Guatemalan Raspberries

Like *Cryptosporidium, Cyclospora* is a parasitic protozoan transmitted through feces. It infects human beings' small intestine and usually causes watery diarrhea, stomach cramps, nausea, vomiting, tiredness, and low-grade fever. The first human case was diagnosed in 1977, but since 1996 numerous cases have appeared in New York and at least a dozen other states. In the three months from May to July 1996 alone, almost a thousand laboratory-confirmed cases were reported to the CDC. Luckily, no deaths occurred.

The question was, where had the *Cyclospora* come from? As researchers began to trace the germ back to its origin, they found that most of the reported illness stemmed from eating imported produce such as raspberries from Guatemala, mesclun lettuce, and basil. In the case of the Guatemalan raspberries, farmworkers contracted *Cyclospora* by drinking untreated water. Inadequate hygiene was then enough to transmit the germ to the crops they tended and picked. Other routes of transmission may also be involved; *Cyclospora* has now been found in a wide variety of animals, including nonhuman primates such as baboons and monkeys. Much remains to be discovered about this parasite, and that is a telling fact. It is yet another reminder that germs can pop up and surprise us from every corner of the global village we have come to inhabit.

The American food market is increasingly vulnerable to such germ invaders, because of our rapidly growing population's insatiable appetite for the most varied sorts of foods, no matter what the season. Ironically, American farms can grow so much of staple crops such as wheat that enormous quantities go to waste every year, even after the export of millions of tons to other countries. But to satisfy our fickle palates from January to June, we almost must import many other foods such as raspberries. The health costs of that are only beginning to be recognized.

The Coldest Cut of All

Most food-borne bacteria prefer warm, humid conditions, but not all. The bacterium *Listeria monocytogenes,* the cause of listeriosis food poisoning, resists freezing (it is therefore said to be "psychrophilic") and actually thrives in the refrigerator. It can also tolerate a ten percent salt solution, which would be an effective measure against most other food-borne germs.

Ready-to-eat foods such as soft cheese, coleslaw, paté, frankfurters, and cold cuts are the ones most likely to transmit *Listeria monocytogenes.* The longer such food is stored in the refrigerator, the greater the potential problem becomes. Because the germs grow so well in cold temperatures, food kept in the refrigerator for more than ten days can have very high counts of *Listeria*

Listeriosis occurs less frequently than other gastrointestinal diseases. Only about twenty-five thousand cases are reported every year. In the early stages the disease can resemble the flu or mononucleosis. Most healthy people can shrug it off if they get it, but the elderly and members of other high-risk groups can die from it. In pregnant women, *Listeria* bacteria can cause miscarriages and stillbirths. A recent report in the *New England Journal of Medicine* underscored the disease-causing potential of *Listeria,* even in healthy, immunocompetent people. At a large public function, corn salad containing the germ was served to 2,189 persons. Seventy-two percent of them came down with fevers and gastrointestinal distress.

No Free Lunch

Although some of these diarrheal threats seem to have emerged only recently, others have certainly been around since the first mammals evolved. In fact, the human species' susceptibility to diarrheal disease may hold the answer to a long-standing question in evolutionary anthropology. For the last thirty years, anthropologists have been debating claims that the earliest human beings regularly ate carrion, decaying meat scavenged from other predators' kills. Those who believe carrion was regularly eaten base their contention on two things. The first flows from the fact that the earliest human beings and their immediate hominid ancestors must have lived by hunting and gathering, with a rough division of labor between hunting males and gathering females and children. Hunter-gatherer societies survive in a few remote areas of the modern world, and these peoples do scavenge meat from time to time. The wrinkle here is that modern hunter-gatherers have fire and thus can cook scavenged meat, killing off any diarrheal disease-causing pathogens. This begs the question of whether scavenging meat could have been a successful strategy for early human beings before fire was used to cook food.

The second thing that the pro-scavenging theorists rely on is the importance in the human diet of proteins and micronutrients that are found most abundantly in meat. Most of the calories in the diets of modern hunter-gatherers come from gathering fruit, roots, nuts, and other plant foods along with birds' eggs and any small game such as turtles, lizards, and insects that can be picked up along the way. But protein from hunting larger animals is a crucial plus factor, especially for pregnant and nursing women, as it must have been for our remote ancestors. Indeed, regular consumption of some meat protein seems to have played a significant supporting role in the evolution of human beings' large brains. One of the most curious things about human beings is that we have chimpanzee-sized bodies, but King Kong–sized brains. Yes, that's right, if human brain and body sizes had the same relative proportions as exist in all other primates, the average human being would have a body as big as King Kong's. An inescapable consequence of this is that human beings require more high-quality protein during fetal brain development and early childhood than other primates do.

These two factors have persuaded some anthropologists that the earliest human beings could not have passed up a free high-protein meal if it became available through opportunistic or deliberate scavenging. Other anthropologists take the opposite view and maintain that scavenging must have been a very rare practice. It would be wonderful if an archaeological dig could settle the question. But there is no clear surviving physical evidence that early human beings engaged in scavenging. Some archaeological sites have been pointed to as scenes of scavenging behavior, but the evidence from those sites is just as compelling, if not more so, for their being places where hunting occurred. To settle this debate we have to rely on circumstantial evidence and logical inference.

Recently I had a chance to explore this question with Dr. Sonia Ragir, an anthropologist at the College of Staten Island in the City University of New York, and Dr. Martin Rosenberg, a colleague of mine at New York University Medical Center. Together we collaborated on an article for the *Journal of Anthropological Research,* in which we argue that scavenging could not have been a stable strategy for early human beings, or indeed for any hominids or other primates without the means to cook their food. Our argument runs more or less as follows.

In the first place, scavenging is not a very efficient means of procuring food. Even the modern hunter-gatherers who engage in the most scavenging get only a small fraction of necessary proteins and micronutrients from it. Hunting, especially cooperative hunting by groups of males, is much more productive. In this regard it is fascinating to note that human beings are not the only omnivorous hunting-and-gathering primates. Other omnivorous primates such as chimpanzees and baboons also hunt and gather their food. Chimpanzees will gang up on and eat monkeys, for example, although they are much less successful hunters than human hunter-gatherers. Yet nonhuman primates scavenge meat even less frequently than human hunter-gatherers. In one reported instance that Drs. Ragir and Rosenberg and I cite in our article, a troop of approaching chimpanzees frightened a leopard away from a freshly killed bushpig. The chimpanzees would have eaten the pig with gusto if they'd been fortunate enough to kill it themselves. But when they happened upon it as carrion left behind by another predator, they simply looked at it for a while and then moved on to search for other food. The observational data compiled on all nonhuman primates show that they will occasionally eat scavenged meat, but that they never continue doing so for long.

The underlying reason for this, Drs. Ragir and Rosenberg and I believe, is that eating uncooked carrion makes primates sick to their stomachs with vomiting, nausea, abdominal pain, and diarrhea. In fact, persistent scavenging without cooking would sooner or later be a virtually suicidal strategy. The problem is that a dead animal's meat quickly acquires a high bacterial load from the saliva of predators or that of scavenging animals such as hyenas, lions, and wolves, from the digestive tracts of insects, and from its own gut. Most predators open their kill through the soft underbelly and eat the visceral organs first; this insures that intestinal germs will spread throughout the prey animal's flesh. The accumulation of diarrheal-disease-related germs in carrion accelerates in warm tropical and subtropical climates like those of the African regions where *Homo sapiens* evolved from ancestral species such as *Homo habilis* and *Homo erectus*. The bacterial load of carrion can reach dangerous levels within an hour or two in such circumstances.

Carnivorous predators and scavengers, such as hyenas, vultures, lions, and jackals, get around this problem because of their gut morphology, that is, the structure of their digestive systems. The gut morphology of predators and scavengers enables them to hold large quantities of food in their stomachs for long periods of time, during which the food is repeatedly bathed in germ-killing stomach acid. Only a little food at a time is released into the intestine, which it then passes through very quickly. Any disease-causing germs that survive prolonged exposure to the stomach's harsh acids have little chance to lodge in the intestine and start a debilitating infection. The upshot is that predators and scavengers can gorge on large quantities of meat at one time, and also return to a cached kill for later meals, as animals like leopards often do, without fear that the decaying meat will make them sick.

Omnivores like human beings and chimpanzees, two species that devote endless creativity to finding new things to eat and new ways to eat them, must employ exactly the opposite digestive strategy. In many respects it makes more sense to think of the hunting and gathering behavior of early human beings in terms of chimpanzee behavior than in terms of the behavior of present-day hunter-gatherer societies. Early human beings were like chimpanzees in the structure of their guts, their appetite for many different kinds of foods, and their lack of fire and ignorance of cooking. (Only the last analogy no longer holds true today, but then again perhaps we shouldn't be too sure we know exactly what our chimpanzee cousins do and don't know! Maybe the female chimps are too smart to get caught doing all the cooking, as well as most of the child rearing.)

The diet of early human beings must have been very like that of chimpanzees today: fruit supplemented with herbs, roots, nuts, seeds, and animal and/or insect protein. To digest this varied diet, especially tough plant matter, omnivores hold food in their stomachs for only a short time before passing it into the intestines, where enzymes released by the pancreas and the normal gut flora break it down in a complex process of fermentation. Food may pass out of the stomach as quickly as thirty minutes after it is eaten, and then remain in the intestines for up to forty-eight hours. This means that the intestines suffer prolonged exposure to whatever germs survive a minimal acid wash in the stomach. As a result, omnivorous primates like human beings and chimpanzees are much more susceptible than scavengers and carnivorous predators are to the debilitating, potentially even life-threatening diarrheal disorders that occur when infectious germs release toxins that irritate the intestine.

The intestines are not defenseless. Pancreatic enzymes can break down small amounts of *Clostridium perfringens* toxins, for example, into harmless amino acids. This could have allowed early hunter-gatherers to ingest the intestinal germs in insects and perhaps even small reptiles and mammals without too much distress. But remember the Theobald Smith equation that we looked at in Chapter 5, $D=NxV/H$, where D is disease, N the number of germs, V their virulence, and H the host environment. If there are too many toxin-producing germs in a primate's food, its natural body defenses will be overwhelmed.

A primate's digestive system can be overwhelmed simply by eating too much meat at one time, even if it is cooked, if the prior diet has been low in protein. A low-protein diet does not elicit much activity from pancreatic enzymes, which means that gorging on meat can surprise and overload the system. This sometimes creates trouble for the indigenous peoples of the remote highlands of Papua New Guinea, who generally eat a low-protein diet like that of ancient hunter-gatherers. The New Guinea highlanders celebrate great occasions, however, with elaborate pig feasts, during which they consume massive quantities of pig meat, yams, and sweet potatoes that have been cooked over an open fire. Complex carbohydrates such as yams and sweet potatoes contain protease inhibitors that are deactivated by thorough cooking. But open-fire cooking often leaves the core of the yams and sweet potatoes relatively untouched, which allows still active protease inhibitors to deactivate digestive enzymes that would ordinarily destroy *C. perfringens* toxins present in meat. Whole villages fall ill as a result, and children between two and four years of age often die. (Younger children are protected because they are still being nursed by their mothers, whereas older ones have more mature immune systems.) The highlanders call this kind of sickness Pig Bel; its formal name is *Enteritis necroticans*.

To sum up the theory that Drs. Ragir and Rosenberg and I have advanced then, early human beings would have had strong incentives to avoid carrion. No doubt the experiment of eating carrion was tried again and again over the course of early human history, especially in times of scarcity, but not by the same individuals. A few trials would be enough to convince any primate as intelligent as early human beings or chimpanzees that an apparently free lunch had extreme hidden costs. For that matter, experiments have shown that it doesn't take a very big or complicated brain for animals to make this association and act accordingly. Known as the Garcia effect for the researcher, J. Garcia, who described it in the 1970s, this phenomenon of behavioral conditioning has been shown to occur in animals as diverse as rats and coyotes. A single trial may be enough to condition an animal to avoid the taste of a food that makes it sick, even if the sickness follows many hours after eating, and it rarely takes more than three to five trials.

In this regard it is interesting to think of the symptoms of gastrointestinal distress, such as nausea and vomiting, not as aspects of infection per se, but as part of the body's effort to fight infection and flush toxins and toxin-producing germs from the system. Scientists have identified a specific chemical that is released by the immune system when it senses a contaminated food. Called tumor necrosis factor, or TNF, it stimulates the brain center for nausea and vomiting.

Garcia-effect experiments have also shown that the opposite is true. That is, a single taste of something which eases gastrointestinal distress will also be remembered afterward and associated with a relief of symptoms. This has great implications for early human beings' trial-and-error discovery of plants and other natural substances that can be used to heal illness. You might say that all of medicine, and especially the modern pharmaceutical industry, rests on a feedback loop that links our taste buds, our digestive systems, and our brains.

Carrion avoidance may have even more profound implications than that. To see what I mean, let's try the thought experiment of going back in time to observe a group of our hominid ancestors as they look for food. To maximize its food gathering, the group often splits into two, with the women and children gathering plant foods and hunting small game, and the men cooperating to hunt larger animals. Suppose we leave the women and children behind for a bit and walk with the men as they hunt. Following a game trail that meanders from savanna to forest edge, the men soon come upon the cached body of a large antelope-like ungulate, apparently killed by a leopard. The meat is still relatively fresh, but the men ignore it because bitter experience has taught them that it will make them sick. They continue the hunt and before long they are lucky enough to kill a fat bushpig. No doubt the men's first step will be to eat their fill of the fresh raw meat. Remember, these ancestors of ours haven't yet learned to use fire to cook.

But what happens next? Bitter experience also tells the men that if they keep gorging on raw pig meat they will overload their systems and make themselves as sick as if they'd eaten the antelope carrion. That leaves two options: They can abandon the rest of their own kill as carrion for scavengers, or they can share it as quickly as possible with the women and children. For me this puts a whole new light on such age-old expressions as "bringing home the bacon." It suggests that the gut morphology of human beings may be a crucial basis for all of our social arrangements, from the nuclear family on out to the modern nation state. As we'll see in a moment, one important way of evaluating a society at any stage of human culture is in terms of how it organizes food production and food sharing. If my colleagues and I are right in saying that gut morphology encouraged carrion avoidance and limited gorging even on fresh meat, then it logically also encouraged the emergence of pair bonding between men and women and the development of the nuclear family as the core social unit!

Why something similar never happened with our chimpanzee cousins is an interesting question. When male chimps succeed in killing a food animal, they sometimes share an odd portion or two with a female. (Although female chimps without offspring sometimes join in the hunting, the activity is primarily a male one.) But this sharing is too sporadic and haphazard to be of much real use to the female or her young offspring; they have to get by on other foods, such as fruit, edible leaves, and insects. The basic difference that enabled different social responses in human beings as compared with chimpanzees may be simply that human beings are much better hunters than other primates. This would have meant that there was more meat to share, and thus a greater incentive to avoid wasting what the males couldn't eat at the kill site.

From Cooking to Pasteurizing to Genetic Engineering

Our excursion into evolutionary digestion, so to speak, has taken us back into our species' distant past. *Homo erectus,* the closest ancestor of *Homo sapiens,* the modern human species, appeared about 1.5 million years ago in Africa. *Homo sapiens* emerged 150,000 years ago, although some researchers think the date is even more recent, about 30,000 years ago. Cooking came along during the transition from *Homo erectus* to *Homo sapiens;* archaeological evidence of widespread use of fire for cooking dates to about 300,000 years ago. But modern human beings' hominid ancestors were successful, tool-using hunters for hundreds of thousands of years before then. They also had some ability to use fire for light, warmth, and perhaps for hunting, just as indigenous peoples of the historical past sometimes set fires to flush prey animals out of their hiding places and make them easier to hunt. These two factors raise the question of why it took so long for people to start cooking their food.

I don't want to beat a dead horse too hard, but it is possible that carrion avoidance played a role in this time lag. It is theorized that people began to cook the meat obtained from hunting after experimenting with eating the flesh of animals killed by forest fires. Early hominids must have encountered such animals countless times before they began to use fire for cooking. Carrion avoidance suggests that hominids would have been very wary of eating found meat, no matter what state it was in. But eventually they must have noticed that burned meat did not decay in the same way as raw meat, and this would have encouraged them to experiment with fire and food. Once they started to do this, they quickly discovered that cooking meat made it tastier, easier to chew and digest, and safer.

For quite a while, cooking meant only roasting spitted meat over an open fire. Evidence of more sophisticated cooking does not appear until about thirty thousand years ago, the time when Neanderthal peoples were disappearing and Cro-Magnon peoples, who were biologically indistinguishable from modern human beings, became the dominant hominid species or subspecies. The first such evidence has been found in southern France, among the archaeological remains of the Aurignacian peoples. These culinary innovators—there must have been many others in other parts of the world, although the champions of French cuisine would doubtless be happy to claim sole priority—wrapped their food in wet leaves and steamed it over hot embers. Thereafter people began to develop techniques such as toasting wild grains on hot rocks and heating liquids in shells, skulls, or hollowed stones. More elaborate cooking required the invention of pottery, which occurred in the Neolithic period, beginning about twelve thousand years ago, roughly at the same time as two other food break-throughs, agriculture and animal husbandry.

The oldest surviving agricultural implements, flint-edged wooden sickles for gathering wild grains, date to 11000 to 8000 B.C. Around the same time, people began to domesticate dogs to use in hunting and, pet lovers will be sad to hear, as food themselves. Sheep and goats were domesticated soon afterward, and by 4000 B.C. people had begun to keep herds of cattle and pigs. Overlapping these developments, from about 6500 to 3000 B.C., people cultivating food plants in the so-called Fertile Crescent area of the Middle East—from the Nile River to the Tigris and Euphrates rivers in present-day northern Egypt, Israel, Lebanon, Syria, Iraq, and western Iran—became the first settled farmers. With agriculture and animal husbandry, humanity assured itself of an ample food supply. These innovations also made possible the rise of complex hierarchical societies like the ancient civilizations of Egypt and Sumer, which arose respectively at the western and eastern ends of the Fertile Crescent.

Humanity had not yet acquired the germ theory of disease, however, and so had few means to thwart food- and water-borne illnesses. Through a long process of trial and error, people came to follow eating habits that then became codified in dietary laws such as those of the ancient Hebrews and early Islam. Ideas that some animals were clean, and therefore safe to eat, and others unclean led to such provisions as bans on eating pork, which must be cooked through to be safe to eat. In many Asian societies, the list of unclean foods included meat from camels, dogs, and human beings, rice left to stand overnight, and anything sullied by rodents. In the tenth century A.D., the Byzantine empire forbade the consumption of blood sausages (a mixture of pig's blood and suet), because so many people died horribly from eating them. The medical experts of the time attributed this to miasmas, but the real cause was botulism toxin.

One of humanity's most pressing problems was how to pre-serve food. For thousands of years, drying and salting were the only means of food preservation, but these techniques, although reasonably effective for meat and fish, do not work well for many other sorts of foods. The problem was only solved at the beginning of the nineteenth century, during the Napoleonic Wars, when a Parisian pastry chef, Nicolas Appert, devised a process for heating foods and sealing them in glass jars. Appert maintained that this destroyed the disease-causing "ferment" in foods. The French navy adopted Appert's products, and in 1809 Napoleon rewarded him with a prize of twelve thousand francs. Appert even wrote a book on his invention, *The Art of Preserving All Kinds of Animal and Vegetable Substances for Several Years,* which was translated into English in 1811. Sadly, Appert's factory was destroyed and he died in poverty in 1841, after a coalition of British, Prussian, Russian, and Swedish troops reached Paris and deposed Napoleon.

Half a century later, Louis Pasteur came to the problem of food safety from his studies of bacterial fermentation in beer and wine. He put Appert's heat-preservation method on a scientific basis, and it became known by his name, as "pasteurization." Today countless foods and drinks, such as cheese, milk, and juices, are pasteurized so that they can be packaged safely and sold over long periods of time. In the century and a half since Pasteur's time, subsequent developments in food safety, preservation, and packaging—refrigeration, advanced drying and freeze processing techniques, and chemical preservatives—have produced results that can be seen in every modern kitchen: refrigerators full of fresh produce, meat, eggs, cheese, and milk; cabinets stocked with canned and packaged goods; and more of everything whenever we want from the corner store or the local supermarket.

In the industrialized world these developments have been accompanied by government regulations that help to insure that the things we eat and drink will not hurt us. But as I noted in connection with the Guatemalan raspberries, more and more foods are being bought and sold across international boundaries without any common standards of food safety and hygiene. As we'll see shortly, there is also the problem of enforcing regulations when they do exist, even in technologically advanced countries like those of North America and Western Europe.

Complicating matters even more are the promise and the threat of genetic engineering. The promise lies in genetically engineered crops such as "golden rice," which contains nutrients that could save countless lives in the Third World. The threat many people reasonably fear is that such innovations will disrupt natural checks and balances in the macro- and microecology of the planet. Genetic engineering is a very powerful tool which must be used with as much care and precision as a surgeon's scalpel. Otherwise, there could be significant unwanted long-term consequences. Some surprises may be in store for us even now, given that one-fourth of U.S.-grown crops are already genetically engineered in some way, and worldwide more than one hundred million acres of farmland are planted with genetically modified crops. Hopefully, we will learn to work with "mother nature" and not against her. If we can do this, then genetic engineering may prove its worth by leading to disease-resistant crops and food animals that might eventually eradicate food-borne illnesses for the first time in human history.

In the meantime, we all still have to practice basic food safety both as individuals and as societies. Unfortunately, many Americans do not know how to protect themselves and their families even in their own kitchens. For a segment on the Montel Williams show, I devised a demonstration to illustrate this. After having his hands coated with a fluorescent chemical, Montel prepared a luncheon meal in front of his studio audience just as millions of Americans do every day; in the process, he violated every conceivable dictum of food safety. For example, he handled both raw meats and vegetables together; he used the same utensils and cutting board for both meats and vegetables; he either did not wash his hands or inadequately washed them without soap; he touched canned goods, countertops, the refrigerator, kitchen cabinets, his apron, and various containers and utensils with unwashed hands; and so on. Then the regular studio lights were extinguished and black lights were turned on to reveal glowing fluorescent dye, which stood in for germs, on virtually every surface and everything in the studio kitchen.

Simple hand washing would have dramatically reduced, if not entirely eliminated, the possibility of contamination in Montel's lunch. But sometimes an individual is at the mercy of larger forces.

Mad Cows and Englishmen

"Mad dogs and Englishmen go out in the midday sun," Noël Coward wrote and sang. If he were alive today, he might well be tempted to change his lyric. Ever since mad cow disease first became headline news, in 1996, concern about it has been escalating steadily, and with good reason.

Mad cow disease is not just a disease of cows. The same disease has long occurred in goats and sheep, in which case it is called scrapie, and in human beings, in which case it is called Creutzfeldt-Jakob disease (CJD) or kuru. Shepherds noted scrapie in sheep and goats more than two hundred years ago, and New Guinea highlanders have traditionally been vulnerable to kuru. Only recently have these disease variants been linked to the same source, an infectious subviral protein particle called a prion. The prion was originally designated a "slow virus," although it is not really a virus, because it ordinarily has the longest incubation period of any known infectious germ, about thirty to forty years. Once symptoms appear, however, the end is near. In whatever creature it strikes, prion disease begins with a shiver and then progresses with jerks and tremors to total loss of motor and neurological control, leading inexorably to death within a year's time. To contract it is to receive a death sentence: There is no known case of remission or recovery once symptoms begin. Interestingly, eating scrapie-infected sheep cannot transmit prion disease directly to human beings. Something happens to the prions inside cows—and also deer and elk—that makes them potentially dangerous to human beings as well.

What happens has something to do with the way the prion is folded. Prions are actually a normal component of the human body, but in prion disease the prions are folded in an abnormal way that makes them disease triggers. In human prion disease, a genetic susceptibility also plays a role. About forty percent of the British population, for example, have an abnormality in their own prion gene that makes them susceptible to the strangely folded prions found in mad cow disease. These patients possess the specific genotype "MM" as opposed to the other two possible genotypes, "MV" and "VV." Since only about one hundred people in Great Britain became sick from ingesting material from mad cows, some co-factor must also be involved. Professor Stephen DeArmond of the University of California at San Francisco has suggested that bouts of tonsillitis may facilitate transfers of the abnormally folded prions. The tonsils are known reservoirs of prions, and prions in the tonsils can easily migrate into adjacent lymph nodes. This would also explain why most of the victims have been relatively young, because younger people are most susceptible to repeated bouts of tonsillitis.

The prion is uniquely different from every other known germ in its structure and effects. It is not even a full cell, and unlike viruses it contains no nucleic acid but is made up solely of proteins. In the early stages of infection, prions can be found in low levels throughout the bloodstream and in body tissues, but over time they concentrate themselves in the brain, eyes, and spinal cord, and these are the only parts of the body they damage. They particularly attack the brain's axons and neurons, eating them away to create the spongelike appearance noted by the researchers who gave mad cow disease its formal name of bovine spongiform encephalopathy (BSE). This manifests itself in disturbed behavior on the part of the infected animals. Scrapie got its name because infected sheep and goats obsessively scrape their sides against trees, posts, and any other suitable object. Mad cow disease got its name in 1986, when a British farmer called his vet to report that one of his cows was "dancing."

Prion infection elicits no immune response from the body, which apparently does not even recognize the abnormally folded prions as foreign substances. They do not inflame body tissues, and they trigger no responses from B or T cells, the immune-system cells that seek to bind with and engulf every other known infectious germ. To top things off, prions are hard to kill. They resist standard methods of decontamination, but they are sensitive to a few things: a ninety percent phenol solution, autoclaving (killing germs by using steam under pressure), incineration, pure lye, and concentrated household bleach.

Medical science recognized the first human cases of this horrible disease in 1977 among the peoples of the New Guinea highlands, who have long known it as kuru. The key to the nature of the disease came when researchers linked it to traditional funeral practices, which involved ritual eating of the dead as an act of mourning and a final affirmation of kinship. When the New Guineans stopped eating their dead, they stopped the disease.

In 1986, a similar disease process was identified in British cattle. By 1993 the documented tally of diseased cattle crested the 100,000 mark. It was determined that the cows were being infected because their feed contained ground-up brains of scrapie-infected sheep, and then in quick addition those of BSE-infected cattle, as well, as a protein supplement. Later it was also determined that maternal transmission can occur during an animal's pregnancy, as prions migrate into the womb to infect a developing fetus.

With regard to human prion disease, or CJD, the thought of prions' eating holes in your brain after thirty or forty years is frightening enough, surely. But in 1996 CJD began to kill British people in their late teens and early twenties. When these cases of variant CJD (vCJD) were linked to eating British beef, the mad cow scare began in earnest. "You eat it. Then it eats you" was the message the media shrieked into the public's ear. Meanwhile the British authorities reacted slowly, delaying the implementation of a necessary temporary ban on the sale of British beef. Even after the ban went into effect, including a ban on the domestic sale of feed containing animal parts, the British government allowed that same feed to be exported to other countries, including the United States.

The British Ministry of Agriculture eventually had to order the destruction of over four million British cattle as part of an ongoing effort to eradicate the disease. To date 200,000 British cattle have died of BSE, and the toll will continue to rise. The beef industry and cattle growers have suffered catastrophic financial losses. It is understandable that politicians fear actions that threaten a country's livestock industry. But delay only makes matters worse for everyone, including the beef business. The cost in any additional lives lost is inexcusable. Since 1996, when the faster-acting variant of CJD appeared in Great Britain, about one hundred people have died of it there, two in France, and one in Ireland. Some experts have predicted that the death toll in Great Britain could reach 500,000 people over the course of the next thirty years.

Unfortunately, a pattern of confusion and delay has been repeated so far in every other European country where mad cow disease has occurred. The results are a stark warning to America, Canada, Mexico, Australia, and other countries where mad cow disease has not yet appeared, that they must be vigilantly on guard to prevent its entering the food chain. They must also be ready to take appropriate measures immediately, if and when it does.

Many efforts to solve the mysteries of prion disease are now under way. (Preliminary research indicates that anti-prion antibody as well as some drugs such as quinacrine and chlorpromazine show therapeutic promise.) Some leading scientists have proposed that there might be a third mode of infection besides maternal transmission or eating the brains, spinal cords, or flesh of infected animals. The theory is that cow feces might contain prions and that other cattle might ingest them, or fecally contaminated soil, when they are pastured or penned together. There is as yet no evidence to support this theory, but there is evidence for something equally disturbing.

In an experiment with laboratory rats, scientists have discovered that animals may develop a subclinical form of prion disease. Although they carry lethal levels of prions in their brains, they have no symptoms. Yet prions from their bodies can produce a lethal form of the disease in other animals. In this regard it is worth noting that prion disease can affect a wide spectrum of animals, from poultry and pigs to cats and dogs. In November 2000, British zookeepers had to put down a lion that contracted prion disease from being fed whole carcasses, including the brains and spinal cords, a practice that British zoos have now banned. This makes me wonder about the safety of dog food that contains the whole gamut of animal parts!

To date no case of prion disease has been documented in American cattle. Nor has there been any reported incidence of a faster-acting variant of CJD in the American population. In fact, the U.S. rates of normal, so to speak, CJD have been stable since 1979, at around one in a million cases per year, which is certainly encouraging. On the other hand, there have been several reports in recent years that elk, deer, and moose in the western United States and Canada are dying of chronic wasting disease (CWD), which is apparently identical with prion disease in cattle. It has been speculated that the disease got into these animal populations because of contaminated British feed sold to North American game farms. Alternatively, horizontal transfer could have occurred from sheep to deer at the Foothills Research Station Confinement Facility at Fort Collins, Colorado. A herd of eighty-one elk in Montana had to be destroyed in 2000 because nine of them had CWD. The possibility thus exists that hunters and others will contract prion disease from eating the meat of infected game. In addition, it has been shown that the prions involved in CWD can be transferable to cattle. This creates another possible avenue by which mad cow disease might break out in the United States.

For the present we can count ourselves lucky that prion disease has not appeared in American livestock, and we can still enjoy eating beef. But this situation can only continue if we are vigilant. Consumer watchdog organizations, such as the Public Citizen Health Research Group, warn that the U.S. may not have adequate safeguards in place to keep mad cow and other animal diseases from contaminating the food chain. In January 2001, the Food and Drug Administration admitted publicly that the American feed industry was not following rules designed to keep animal brains and spinal cords out of the food chain. Later that same month, the Texas Animal Health Commission quarantined a thousand cattle as a precautionary measure, when it was found that their feed might have contained ground-up bones, brains, and spinal cords. According to the federal government's General Accounting Office, the majority of companies in the U.S. cattle industry, perhaps seventy percent of the total, do not comply with the existing rules. If we adopt too laissez-faire an attitude on this sort of thing, many innocent Americans could die, and that would damage the American beef industry, as well. Food safety is an area where appropriate commonsense regulations and enforcement are in everyone's interest.

I suspect that prion disease may not threaten American beef eaters so much as it may American blood bank users. A possibility exists that blood donors who were resident in Great Britain during the 1980s and 1990s could pass prions into the American blood supply. This mode of transmission has been shown to be feasible in animal experiments. Blood banks have therefore had to begin asking prospective donors not just about their past and current health, but about whether they spent any substantial period of time in Great Britain during those two decades. If they've lived three months in Britain, six months in the UK, or if they've received a transfusion there at any time since 1980, they cannot donate blood. For their part, British blood banks now take only the top layer of donated blood, the plasma portion, shedding both the middle layer, the so-called buffy coat component, which contains white blood cells, and the heavy bottom layer, which contains red blood cells. The buffy coat component of the blood carries the highest risk for prion transmission. The central puzzle of prion disease lies precisely here, in prions' ability to infiltrate the buffy coat component without triggering an immune response from the white blood cells, and specifically from the B and T cells.

Protecting Our Food and Water

The number of food- and water-borne germs that threaten us with illness can seem very daunting. It is important to look at these threats in two contexts. One is how deadly an infectious germ is if it is contracted. This is where we tend to focus our attention. But the other, equally important context is how likely a person is to contract a particular germ in the first place. The deadliest germs almost always tend to have a low incidence in a population, because the quicker a host organism dies, the less opportunity the germs have to migrate into new hosts. Usually such germs need human behavior—individual, social, industrial, or governmental— to give them a helping hand.

So although prion disease inevitably kills anyone it infects, it cannot infect many people without a big boost from human negligence or error. Potentially lethal diarrheal disease will always remain a much greater killer in terms of sheer numbers, because of the variety of germs that can cause it and the relative ease with which they can contaminate food and water supplies. The bottom line is that we can contain the impact of both sorts of threats, however, if we exercise common sense as individuals and societies. The risks to health are real, but they are manageable. We do not need to live in fearful anxiety about what we and our families are eating and drinking. If we are armed with the right Protective Response Strategies, we can do a great deal to insure that our food and drink will nourish rather than sicken us, we can continue to enjoy the mouthwatering variety of the world's cuisines, and we can have a positive influence on the safety of the food and water in our communities.

Protective Response Strategies for food and drink include the following:

- Before you start preparing a meal, take a moment to review the foods you're going to use. This will lessen the chance that a moment's rush or confusion will make you forget proper cleanliness. Creative cooks need not fear losing their spontaneity; even a rough idea of what's ahead can help keep a healthy, delicious meal on track.

- Wash your hands often with soap and water (see pages 28–29). Wash in between preparation of different foods and after completing different stages of preparation; after throwing away debris; after greeting guests with a handshake; before setting the table; and after using the bathroom. Also remember to wash your hands after touching inanimate objects such as doorknobs, latches, or pieces of furniture. If you are eating in a public place and must visit the rest room, use paper towels to exit the toilet to prevent reinoculating yourself with germs.

- Cut meats and vegetables with separate knives and cutting boards, or carefully wash knife and cutting board in between the two. Look for plastic cutting boards with an embedded antibacterial substance such as triclosan. Such a product cleanses the inner surfaces of slices and fissures in the board.

- Cook meats and poultry thoroughly. Eighty to one hundred percent of all poultry contain gastroenteritis-causing *Salmonella, Campylobacter,* or both. Never eat poultry if it is pink on the inside. Chicken and other poultry should be cooked to an interior temperature of at least 180 degrees Fahrenheit. The same holds true for fish, which can carry the diarrhea-causing germ *Vibrio.* Steak can be cooked rare on the inside, because germs can only reach the outer layer of the meat. Hot dogs and hamburgers, however, which may contain ground-up bone and other animal parts, should also be cooked to an interior temperature of at least 180 degrees Fahrenheit. Make it a habit never to eat any raw meat, poultry, or fish. And never eat raw eggs unless they are pasteurized.

• "Never eat raw fish?" outraged sushi lovers will cry. Well, I can only say that I don't care to take the risk. But I know many who can't resist, and sooner or later they must be prepared to accept some gastrointestinal distress from a *Vibrio* germ. Worse, they could ingest the parasitic anisakis worm. Sushi can be perfectly safe, but as with many other foods, it all depends on how the fish is selected and handled. An expert sushi chef will be able to tell from its appearance and odor whether a fish is of the right quality, and often he can remove the anisakis parasite without spreading the germs to the rest of the fish, a very nifty bit of surgery. When you eat sushi, you're really in the sushi chef's hands. Since expertise varies from chef to chef, you have to prepare yourself for a bout of illness now and then.

• In the case of shellfish, remember that freezing for twenty-four hours beforehand will not kill all potential pathogens. Cooking is the way to go. Domestic shellfish and bay-caught fish face growing risks of infection from agricultural runoff and from discharged ballast water. Researchers at the Smithsonian's Environmental Research Center found that a gallon of the ballast water that foreign ships typically discharge into the Chesapeake Bay contains about three billion bacteria, including cholera bacteria, about twenty-five billion virus particles, and unknown quantities of other marine life. Foreign ships empty about 2.5 billion gallons of ballast water into the Chesapeake Bay every year. This additional germ load must be putting great additional stress on the bay and its natural life, which is already under attack from many pollutants, and it could be introducing numerous nonindigenous waterborne diseases into the ecosystem and thus into the food chain. The same thing of course happens in other bays, harbors, and ports on the American coastlines. Remember, *Vibrio* species and other food-borne pathogens are no laughing matter; they can be potential killers of the immunocompromised and the elderly.

- Assume that the outer surfaces of all fruits and vegetables are contaminated. Remember that imported produce may not have been grown or harvested in sanitary conditions, and avoid it whenever possible.

- Clean fruits and vegetables carefully. Soak produce for five to ten minutes in a room-temperature solution of equal parts water, hydrogen peroxide, and vinegar or lemon juice or ascorbic acid (vitamin C). Then rinse well to remove any debris. Scald raw vegetables like carrots and celery in hot water whenever possible to remove the outermost layer as well as any rotted portions. A veggie brush is an indispensable utensil for cleaning all hard vegetables such as carrots, beets, potatoes, radishes, parsnips, and turnips; every kitchen should have one.

- Resist the temptation to sample unpasteurized cheeses and beverages, such as apple cider or orange juice. Pasteurization helps kill diarrhea-causing *E. coli* 0157:H7 and *Salmonella*. Always serve children pasteurized juices and foods.

- Don't leave food sitting out at room or outdoor temperature for an extended period of time during cocktail parties, family gatherings, and barbecues. Food should be brought out less than one hour before guests arrive and should be left out for no more than two hours in a warm environment. Mayonnaise-laden salads and other creamy foods, for example, provide a hospitable growth medium for *Staphylococcus aureus* germs that are part of many people's normal skin flora; as we saw earlier in the book, *S. aureus* toxins can produce food poisoning.

- Also apply the two-hour rule to leftover restaurant food that has been wrapped up to go. Such food could have potentially high bacterial counts and thereby present an unnecessary risk. Refrigerate restaurant leftovers as quickly as possible, and reheat to a temperature of 165 degrees Fahrenheit before re-serving.

- Reheat and serve leftovers only once. Leftover food will have a higher bacterial count, and some bacteria can grow very well at refrigeration temperatures (about 40 degrees Fahrenheit).

- Never store cold cuts and other ready-to-eat foods such as soft cheeses, coleslaw, paté, and hot dogs for more than a few days to a week at most, to prevent the overgrowth of *Listeria monocytogenes,* which thrives at refrigeration temperature. Pregnant women should avoid such foods.

- Thoroughly cook foods likely to harbor *Listeria,* because these bacteria are not killed by pasteurization.

- Pay attention to expiration dates. Always check the date on meat, dairy products, and pre-made salads. The older the food, the likelier it is to carry a large quantity of germs. Ingesting very large numbers of some organisms will cause illness, when a smaller amount would have been no trouble at all. Unfortunately, the poor are at greatest risk here. Cut-rate grocery stores often sell packaged foods long past their expiration date. This food may have a very high bacterial count. For example, hot dogs, for which the usual acceptable germ count is one million per gram, may have counts over 330 million per gram if their expiration date has passed. Apparently the legal onus is on the consumer. It is truly a question of buyer beware.

- Exercise care with the communal plate. During parties, people tend to take second helpings from large serving plates with their own utensils, which have been contaminated with germs from the skin and mouth. Or they use their fingers. In this way the platters become inoculated with salivary and skin germs, which can include streptococcal and staphylococcal bacteria, as well as with cold, flu, and stomach viruses that are transiently on the fingers. Bacterial germs can multiply very rapidly in food at room temperature. Always supply and use separate serving utensils and discourage guests from using their own contaminated utensils.

- Use antibacterial soap and water if time doesn't permit proper washing with ordinary soap. A small bottle of waterless antibacterial hand rinse can easily be carried in a pocket or purse and used discreetly whenever needed, from street fairs to restaurants with poor toilet facilities.

- If you experience vomiting, nausea, stomach cramps, fever, or headache within twenty-four hours or so of a meal, don't let it go, see your doctor. Even nonfatal foodborne illness can wreak havoc on your gastrointestinal system without proper treatment.

- The label "organic" does not necessarily mean "safe" or "germ free." Organic produce can harbor pathogens just as easily as conventionally grown foods. Carefully soak, wash, and rinse all produce as explained above.

- Never drink or serve water of unknown quality. Drink bottled water, or boil suspect water for ten minutes before drinking.

- The immunosuppressed, the immunocompromised, the chronically ill, young children, and the elderly should drink bottled or boiled water, if the public water supply is from open surface sources like lakes and reservoirs.

- Do not eat imported meat or meat products, especially those from countries where mad cow disease has been found. There is some chance that cheese from these countries could also carry the infectious prions, but they would tend to be present only in very low concentration. Some French wines are clarified with bovine blood, and even some foreign brands of canned tuna may contain bovine products, so you may wish to be careful of these, as well. At present, American meats and cheeses are considered safe to eat.

- Support the adoption of measures that will protect the environment at home and abroad, and thus the global food chain that we all rely on. Let your elected representatives know that you support appropriate food-safety regulations. As I said above, this is one area where everybody, Republicans and Democrats, left, right, and center, should be able to agree that sound, effective regulations are in the interest of all.

- Consider that irradiated food products might finally prove to be useful as a means of extending the shelf life of foods, as well as killing potential pathogens.

- Support initiatives to bring clean water and sanitation to developing countries. In a globally ever more connected world, the health of all the world's people becomes ever more connected. Remember, the diarrheal illness that a Third World farmworker experienced last week may visit your house tomorrow.

8

In Thin Air

The New Assassins

In the summer of 1976, a new killer germ seemed to materialize out of thin air. About two hundred American Legion members attending a convention at a Philadelphia hotel were suddenly stricken with a strange new disease that would claim the lives of twenty-nine of them. The symptoms looked like pneumonia, which has long been known to be caused by a number of different microorganisms, including bacteria, fungi, and viruses. But the new disease spread instantly and had no detectable cause, a combination that created a panic.

The public's fear was not allayed when scientists finally identified the cause as a previously unknown bacterium, and named it *Legionella pneumophila* for its victims and its major symptom. Where had this germ come from? How could medical science have failed to detect it before? Did it spread contagiously, from person to person, or in some other manner? The first dramatic challenge to people's faith in medicine since wonder drugs and vaccines had vanquished diseases such as smallpox and polio in the 1950s, Legionnaires' disease triggered a new anxiety about germs. That anxiety only became more pervasive as the 1980s and 1990s brought news of other previously unknown killer germs, such as those which cause AIDS and Ebola fever.

We now know that people don't catch Legionnaires' disease from one another like the pneumonic plague or tuberculosis, which spreads through close person-to-person contact as infected people exhale germ-bearing droplets from their lungs. *Legionella* bacteria were inhaled by the unfortunate conventioneers, but how those bacteria came to be present in the air was discovered only by brilliant scientific sleuthing. Water-loving germs, *Legionella* live in insignificant numbers in most lakes, streams, and ponds, and even in some grocery-store vegetable misters and household showerheads. Unlike most other bacteria, they can survive at room temperature in plain distilled water for as long as 140 days, and in tap water for over a year. In fact, they can grow in tap water, feeding on the cysteine in the carcasses of dead bacteria and on iron from pipes. It takes many gallons of water to find *Legionella* in any numbers, however.

Because they have branched-chain fatty acids, like the thermophilic archaebacteria that were probably among the first living creatures on Earth, *Legionella* take heat well enough to thrive in hot-water boiler tanks. They can withstand pasteurization temperatures of 65 degrees centigrade, about 150 degrees Fahrenheit. Hot or cold, they can survive being sprayed in aerosolized droplets as far as six hundred feet, which is farther than the longest home run on record. They escaped detection before 1976, despite being present in low concentrations in many open-water sources, for two main reasons. Because they feed only on cysteine and iron, they do not grow in ordinary laboratory cultures. And they are very hard to see under the microscope, resisting most microscopic stains except silver. Deposited on *Legionella* in very thin layers, silver acts like the bandages covering the Invisible Man and allows the shadow of the bacterial cell to be seen.

Legionella bacteria can also survive in cooling towers and, as it happened, in the open rooftop water-supply troughs of the Philadelphia hotel's air-conditioning system. As part of a conservation effort, the water in these troughs, which were located right by the air-conditioning intake duct, constantly recirculated over the system's cooling coils. This gradually increased the ambient temperature of the troughs to the point where airborne blue-green algae that dropped into the water could thrive. Sunlight on the open troughs of warm water promoted rapid algal growth, providing an excellent food source for *Legionella,* which in turn propagated wildly, reaching a population density that would never be found in nature. Association with fresh-water amoebae, often found in potable water supplies, can also enhance their growth. As the *Legionella*-saturated water evaporated, droplets of its vapor were sucked into the building's ventilation ducts. Like huge aerosol sprayers, these ducts spewed the germ-ridden brew everywhere in the hotel.

Today rooftop air-conditioning systems use closed water troughs, and office and service buildings, such as hospitals and hotels, have their water systems checked periodically for *Legionella*. If it is found, the water systems must be purged and decontaminated with superheated water. Even so, immunocompromised patients, such as those with AIDS, have been known to contract Legionnaires' disease from inhaling germ-bearing droplets when they shower in the hospital, where showerheads haven't traditionally been cleaned frequently. Likewise supermarkets have shifted to direct spray systems, replacing misters with reservoirs where *Legionella* could breed, in order to prevent immunosuppressed people like the elderly from inhaling the bacteria. These actions form part of a good community **Protective Response Strategy**, which in addition should include ozonation and microfiltration processes to check *Legionella* and other germ growth in municipal water supplies. Bottled-water manufacturers should take the same precautions.

A personal **Protective Response Strategy** against Legionnaires' disease should be twofold. First, make sure that the water in household misters and humidifiers is always fresh, and clean the misters and humidifiers after each use. Second, remove showerheads once a year and clean them manually with a wire brush to root out any *Legionella* that might be lurking in them. People whose work takes them to construction sites should be aware that *Legionella* may breed in large standing puddles.

In one sense, a new disease did materialize out of thin air at that ill-fated American Legion convention. But this was not because a totally new germ had appeared. It was instead because human action had created a new microenvironment for an age-old germ, which in turn opened up a new avenue of infection. Before very recent times, *Legionella* would not often have accumulated in water in high enough concentrations to be dangerous to anyone, and if it had it would have lacked an aerosolizing delivery device to spread germ-laden droplets into people's lungs efficiently.

Innumerable germs—either entirely unknown to science or not yet identified as playing a role in disease—could be waiting for similar opportunities to break out of their natural obscurity and spread among human beings. Whenever this occurs, medical science must play a frantic game of catch-up to identify the germs involved as quickly as possible and begin to lessen their impact on human health. If the growing scientific consensus on global warming is correct, for example, human-induced climate changes may unleash a new array of infectious germ threats in the decades to come.

Something of this sort may already have occurred in connection with a 1993 outbreak in New Mexico of hantavirus disease, a rare, deadly, incurable infection caused by inhaling airborne particles of rodent excreta. Thanks to fluctuations in the El Niño–La Niña weather cycle, which many researchers believe global warming exacerbates, there was an unusual burgeoning of grasses and other plant life in the southwestern United States that year. This promoted a population explosion among deer mice, which likely have long carried hantavirus, increasing the odds that human beings would be infected. Fifty-three people contracted the virus, and thirty-two of them died, most within days of the onset of symptoms.

Less than a year later, in January 1994, hantavirus struck at the other end of the country, in Rhode Island. A previously healthy twenty-two-year-old man died suddenly of acute respiratory distress syndrome (ARDS), when his lungs failed within hours of his admission to a hospital emergency room. The young man had developed flu-like symptoms of fever and muscle ache only three days before. Subsequent analysis of blood and autopsy specimens by the CDC confirmed that he had died of a hantaviral infection. During the two months preceding his illness, the victim had traveled only to Rhode Island, where he was a college student, to his home on Long Island, and to a family-owned warehouse in Queens in New York City. Because this was the first documented case of hantavirus illness in the northeastern United States, the CDC investigated it in conjunction with the New York City, Nassau-Suffolk County, New York State, and Rhode Island State departments of health. Unfortunately, the investigators were not able to determine how the young man had contracted hantavirus.

Research has shown that rodents with hantavirus shed infectious viral particles in their saliva, urine, and presumably their feces for weeks or months, and that it takes very little contact for human beings to become infected. Laboratory workers have contracted hantavirus after only a brief exposure to caged infected rodents. It has now been determined that hantavirus infects human beings in diverse geographical locations throughout the United States and the rest of the world. The human target organ is usually the kidneys or the lungs. The kidney, or renal, type is called hemorrhagic fever with renal syndrome (HFRS). The lung, or pulmonary, type is referred to as hantavirus pulmonary syndrome (HPS). Death often occurs within only one or two days of infection. In both HFRS and HPS, the initial symptoms mimic those of flu.

Between 1994 and 2000, illnesses meeting the current case definition of HPS were reported to the CDC from sixteen states in nearly seventy people ranging in age from twelve to sixty-nine years. Thirty-eight of these people died, including the young man in Rhode Island. Although not many people have been affected nationwide, there have been enough cases to make it important for doctors to consider the possibility of hantavirus illness in patients who experience acute respiratory distress that cannot otherwise be explained.

Recently a group of physicians in London, England, retrospectively investigated a mysterious late-fifteenth- to early-sixteenth-century illness known as "the sweating disease." They hypothesized that this old disease was caused by hantavirus. The historical records indicate that the symptoms, method of transmission by rodents, and ecology of the disease fit perfectly. This hypothesis may never be confirmed, owing to the fact that no evidence survives except for contemporary descriptions of the disease. But it is reasonable to infer that environmental changes resulting from rapid growth in population and land use brought Londoners of five hundred years ago into unhealthy proximity with previously wild rodent populations.

As today's population continues to explode and human beings voraciously clear forests and other wild areas for their use, they risk exposure to many new germs. As we've already seen, this is an important part of the story of how Ebola virus and the rapidly mutating AIDS virus migrated from monkeys and chimpanzees into human beings. The AIDS tragedy illustrates that it is impossible to predict how an infectious germ will adapt to human hosts. Before 1996, it was assumed that hantavirus could be transmitted to human beings only by aerosolized rodent excreta. But that year, a terrifying new strain of hantavirus appeared outside Buenos Aires, Argentina. Not only did this strain have a high mortality rate, causing death within twenty-four to forty-eight hours, but it could be transmitted directly from person to person. Since there is no known cure for hantavirus illnesses and no protective vaccine, this new strain could have spread like wildfire and triggered a public-health disaster. Fortunately the Argentine hantavirus strain disappeared as quickly and mysteriously as it appeared. Will we be as lucky the next time?

A disease very like hantavirus illness—fast-acting, deadly, and incurable—appeared in California in June 1999. Over the next year, according to the CDC, a girl aged fourteen and two women, aged thirty and fifty-two, came down with a flu-like illness characterized by fever, headache, and muscle aches. All three then experienced acute respiratory distress syndrome, followed by liver failure with hemorrhage, and all three died within eight weeks of developing the first symptoms. The two women and the teenaged girl had nothing in common except living in California; two were from Southern California and the other was from the San Francisco Bay area. None of them had traveled outside her home area in the month preceding her illness. But interestingly, one of them had cleaned rodent droppings in her home a few weeks before she became sick. With this clue, health authorities reexamined the cases and found evidence of infection with an arenavirus in all three.

Arenaviruses include the African hemorrhagic fever virus that causes Lassa fever. The related virus that caused the deaths in California is called Whitewater Arroyo virus (WWA), for the place in New Mexico where it was first discovered in the early 1990s. At that time the virus was found in the white-throated wood rat *Neotoma albigula*. Before 1999 WWA was not known to cause disease in human beings. Indeed, until then researchers had never considered viral hemorrhagic fevers in human beings to be indigenous to the United States at all.

The explosive population growth the United States is currently experiencing from very high rates of immigration is driving problems of urban sprawl to previously somewhat remote areas of the country, especially in the western United States. Hundreds of square miles of California are being urbanized every year, for example, including the fertile farmland of the Sacramento Valley. Henderson, Nevada, has become the fastest-growing city in the country in an area of more or less arid desert. One result of this has been an increase in human encounters with wildlife, including myriad unfamiliar germs. There is always a chance that a germ will respond to a new presence in the environment by trying to infect it; in this case human beings are the new presence. And as the hantavirus and arenavirus cases show, a germ released from one of its normal reservoirs in a previously sparsely populated part of the country can move, via human beings, to more densely populated areas with surprising speed, sometimes in a matter of months.

Therefore, a **Protective Response Strategy** for new illnesses like hantavirus and arenavirus might well include supporting environmentally conscious measures that reduce population growth, sprawl, and the development of open spaces in your city or state. This will obviously be one of the big political issues of coming years. On a community and neighborhood level you can work for good municipal sanitation and anti-rodent campaigns.

A personal **Protective Response Strategy** against these illnesses calls for knowing how they are transmitted and acting accordingly whenever you come into contact with rodents or see signs of their presence, which will usually be their droppings. Don't begin to clean up rodent droppings, or let someone else do so, without putting on a respiratory mask and gloves. Wet the droppings with disinfectant before removing them. Likewise soak rodent carcasses with disinfectant and then double-bag them for removal.

A Killer Returns

The new (to us) airborne killers have yet to show the astonishing ability to spread among human beings that the age-old one of tuberculosis (TB) has. Tuberculosis has waxed and waned throughout human history, becoming especially dangerous in densely crowded urban environments. It can attack any part of the body, but its most obvious target is the lungs. The illness begins in the lungs with a persistent cough and progresses to fever, night sweats, coughing up blood, and, often, death.

Cases of tuberculosis were recorded as early as 4000 B.C., and some mummies show the spinal deformation that the disease can produce. The ancient Greeks knew it as *phthsis,* "the wasting disease." Among some peoples, it was "the King's Evil" and might be cured, it was thought, by a king's touch. The Maya, the Inca, and the Aztec all confronted tuberculosis in their great cities, as did the ancient Chinese. As Europe urbanized from the mid-seventeenth century on, tuberculosis gathered killing power and became the leading cause of death among adults. In London it was responsible for one-fifth of all deaths between 1650 and 1800. Autopsy records credit it with one-third of all deaths in early-nineteenth century Paris.

Tuberculosis really came into its own in the cities of nineteenth-century Europe and America, as well as in many tropical areas colonized by Europeans at this time. The disease killed people of all classes, but it struck hardest at the new masses of urban poor, whose immune systems were lowered by chronic malnourishment. The popular dread and fascination with tuberculosis throughout the nineteenth century eerily prefigure our own obsession with AIDS. A kind of romance grew up around tuberculosis, "the white plague" that so ate at people's bodies in its late stages that the disease was best known by its other name, "consumption." In 1848 Alexandre Dumas *fils,* as they say, the son of Alexandre Dumas *père,* who wrote *The Three Musketeers,* published a famous novel which he soon adapted for the stage, *The Lady with Camellias,* about a love affair doomed by both social conventions and a fatal case of tuberculosis. The play had legs as a star vehicle for Sarah Bernhardt and other leading actresses into the first decades of the twentieth century; Greta Garbo played the title role in a movie version, *Camille,* that also starred the young Robert Taylor.

Verdi's great opera *La Traviata,* first performed in Venice in 1853, is a remake, if you will, of the play by Dumas. The climax of the opera comes when the consumptive heroine, Violetta, dies in her lover Alfredo's arms. Forty-three years later, in 1896, Puccini drew on the heroines of both Dumas and Verdi for his portrait of Mimi, the consumptive heroine of his opera *La Boheme,* who dies in her lover Rodolfo's arms. Recently the young composer-lyricist Jonathan Larson, who himself died tragically of an aneurysm on the eve of opening night, retold the same story for the world of AIDS in the hit musical *Rent.*

But perhaps Charles Dickens should have the final cultural word on tuberculosis: "a disease which medicine never cured, wealth warded off, or poverty could boast exemption from— which sometimes moves in giant strides, and sometimes at a tardy sluggish pace, but, slow or quick, is ever sure and certain." Dickens's description captures some critical elements that help to explain why tuberculosis was such a dread killer in human history, until the advent of antibiotics, and why it now once more threatens urban populations in the industrial world. In the Third World, tuberculosis has long been the leading cause of death, a distinction it now shares with AIDS; wiping out three million lives a year each, the two diseases are far and away the greatest killers on Earth.

"Slow or quick" is the key observation Dickens makes about tuberculosis, and that is indeed one of the most remarkable aspects of the disease. The disease process begins with inhalation of the tubercle bacillus, *Mycobacterium tuberculosis,* in aerosolized droplets less than five microns in size, which are exhaled from the lungs of infected persons. Although human beings are highly susceptible to infection with tubercle bacilli, they are strongly resistant to developing the disease tuberculosis. Of those people who become infected with tubercle bacilli, fully ninety percent heal and recover without ever developing any significant symptoms. About five percent suffer primary tuberculosis, which may kill them. The remaining five percent acquire a dormant form of the disease, leaving them at risk of potentially lethal reactivation, or secondary, tuberculosis at some point in their lives. The immunosuppression that inevitably accompanies old age may make people in the second group vulnerable to reactivation tuberculosis, and so may the immunocompromise that accompanies diseases as different as AIDS and measles. For example, as many as eight to ten percent of people with HIV who get infected with TB, tubercles will develop primary TB within a year. At any one time about two billion people in the world are infected with the tuberculosis bacterium, and about 300 million of them have tuberculosis disease in primary or secondary form. The World Health Organization (WHO) estimates that ten million new cases of tuberculosis disease occur around the world every year.

The key to this "get you now, get you later" disease process is the way that tubercle bacilli outwit the body's immune system. When a person inhales tubercle bacilli, the germs wind up in the alveolar sacs of the lungs, where resident macrophage cells engulf them. At this point it should be game over, right? No such luck. Ordinarily macrophages release enzymes that kill the germs they engulf. But like some other germs, the tubercle bacilli can deactivate these enzymes and immediately start to reproduce inside the macrophages, which carry them via the lymph nodes and then the bloodstream to every part of the body, until the growth of bacilli breaks the macrophages apart from within. At this point the immune system kicks in with a vengeance on ninety percent of the infections. But five percent of the infections lead to full-blown tuberculosis, and the other five percent go into a dormancy stage that may or may not be broken by the stimulus of later stresses.

Whether in primary or secondary form, tuberculosis can easily become a death sentence. During the course of the disease, as the body is racked by fever and bloody cough, the immune system releases tumor necrosis factor (TNF), also called cachectin; this interferes with lipid metabolism, starving cells of essential nutrients and causing the eaten-away look of consumption. (The same tragic erosion of the human body can be seen in late-stage cancer and AIDS patients.) In fact, the disease tuberculosis is more the result of the destruction caused by the body's own immune system than it is the result of the tubercle bacilli's intrinsic virulence.

Over the course of the twentieth century, tuberculosis began to wane in the industrialized world as economic conditions improved and the majority of people began to live and eat better. It remained a frightening killer, nonetheless. A vaccine offered some protection, but until the antibiotic streptomycin was isolated from a fungus-like bacterium in 1944, there was no cure. Other effective antibiotics were discovered, as well, and it was quickly seen that they worked best in combination. Public-health authorities began to cherish the hope that tuberculosis could be wiped out. From 1953 to the mid-1980s, the rate of tuberculosis cases in the United States fell from eighty-four thousand a year to about twenty-two thousand.

But during the 1980s, Congress and the Reagan and Bush administrations cut funds for TB monitoring, treatment, and research, not with any specific intent to let TB have its way but as part of a general policy of dismantling and weakening government regulations. Unintended consequences soon resulted, helped along by a number of other factors. Although tuberculosis rates fell a little further before the decrease ran out of momentum, they soon rebounded, especially in large cities. New York City's TB rate rose from 1,307 cases in 1978 to almost 4,000 in 1992. In the country as a whole, TB cases bounced back to the twenty-thousand mark in 1990. The increase can be attributed not only to lax or incomplete public-health monitoring at the federal, state, and local level, but also to a surge in the numbers of the poor, and all the attendant health problems of poverty, because of increasingly high rates of immigration. The onslaught of AIDS was responsible for many new cases, as well, because its undermining of the immune system left victims susceptible to opportunistic infections of all kinds, including TB. For example, at any one time from twenty-five to fifty percent of all TB patients in New York City are infected with the AIDS virus.

As frightening as the rising number of cases was the high percentage of them that proved resistant to the usual antibiotics. Multi-drug-resistant tuberculosis (MRTB) strains appeared in at least seventeen states. In these states, thirty-four percent of all TB strains were found to be resistant to two or more drugs, and some were resistant to all known antibiotics. MRTB strains proved to be associated with higher than normal death rates of up to sixty percent, and they are particularly heavily concentrated in the AIDS population.

Investigators found that drug-resistant strains of tuberculosis were popping up all over the world mainly because of poor TB patients' failure to complete a full course of treatment. When unsupervised patients stopped taking their medication prematurely, before the combined efforts of the antibiotics and their immune systems knocked out the disease, they created conditions that fostered the growth of antibiotic-resistant subpopulations of tuberculosis bacilli. In response, the WHO recommended a public **Protective Response Strategy** that has proved effective in both America and Africa, a program known as Directly Observed Treatment Shortcourse (DOTS). Under DOTS, TB patients are closely monitored by health-care workers to insure that they complete a course of four drugs lasting six to eight months. The result has been a marked decline in TB deaths wherever DOTS has been implemented; the WHO forecasts that the program will save ten million lives over the next ten years.

However, researchers have found nine hot spots around the world—Estonia, Latvia, Russia, China, India, Iran, Romania, South America, and Africa—where more than three percent of the patients needing treatment for the first time already have drug-resistant tuberculosis. For these areas the WHO recommends a program called DOTSPlus, in which doctors and nurses watch patients take up to seven drugs daily for eighteen to twenty-four months. There is a far greater financial cost per cure, but the new strategy is needed to keep the worldwide tuberculosis epidemic from spinning out of control. Estimates of future tuberculosis infections and deaths worldwide over the next thirty years range from 171 million additional cases with 60 million deaths to 250 million additional cases with 90 million deaths. Even at the low end of that spectrum the consequences will be felt in wealthier nations like America, as well as poorer ones.

Aside from supporting funding for public-health initiatives like DOTS and DOTSPlus, a personal **Protective Response Strategy** for tuberculosis is largely one of being alert to its warning signs and seeking appropriate treatment in a timely fashion. You are unlikely to contract tuberculosis from infected people you spend only a short amount of time with. An occasional ride on the subway, commuter train, or bus with a TB sufferer, for example, will not endanger a healthy, immunocompetent person, whereas daily carpooling with such an individual would heighten the risk of infection dramatically. Just the same, it of course makes sense to move away from strangers who are coughing persistently, or perhaps even coughing up blood.

Sometimes keeping away from coughing individuals is outside your control. As I mentioned in Chapter 4, the CDC has reported the spread of tuberculosis aboard a commercial airliner. The airlines thus need to follow a **Protective Response Strategy** of insuring that ample fresh air is continually circulated through their planes, so that the breath of any passengers infected with TB does not raise concentrations of aerosolized tubercle bacilli in the cabin's atmosphere to a dangerous level. If you find yourself sitting right next to someone who coughs continually, quietly go and ask the flight attendants to reseat you, if possible. If you can't avoid this exposure to another person's germs—even if it's only a bad cold, you have a right to take reasonable action to try to avoid contracting it—be alert to the signs of TB in yourself. If you start to show symptoms such as a persistent cough, a fever that comes and goes, and a loss of appetite and energy, ask your doctor to check it out.

What is far more likely than contracting tuberculosis on an airplane, however, is noticing it in an elderly or ill relative or friend. In January 1999, my mother, then almost eighty-nine, got the flu. After she recovered, she still didn't feel right. Then she began to lose weight, to cough, and to have a recurring fever. At first I dismissed the possibility that it might be tuberculosis, encouraged by my mother, who insisted that all she needed was peace and quiet and for people to stop fussing over her. By the time I got her to her doctor, X rays showed a little bit of fluid around the lungs, along with two very tiny nodules that could have been caused by a long-healed tubercular lesion.

The fluid was aspirated from the sac around the lungs, and a virgin strain of TB was isolated that was not resistant to any antibiotics. This proved that my mother's case was an example of reactivation, or secondary, tuberculosis that she first contracted in her youth. My mother is a first-generation Italian-American who grew up in Brooklyn at a time when poor immigrants were arriving in large numbers and there was a very high rate of TB. Decades later the immunosuppression of the flu, coupled with her advanced age, gave the TB bacilli she had been harboring a chance to rouse themselves from dormancy, which is another good reason for people over fifty to get their flu shots every year. Treatment with vitamin B6 and Rifamate, a combination of the antibiotics rifampicin and isoniazide, stopped my mother's TB in its tracks. She gained her weight and energy back, and sailed into her nineties going strong.

Toxic Atmospheres

The sudden appearance of Legionnaires' disease at a Philadelphia hotel in 1976 impressed the public imagination with the specter of a "sick building" that put everyone in it at risk. Coupled with concerns about the carcinogenic or otherwise toxic qualities of asbestos, polyvinyl compounds, and other materials, this fed a growing anxiety about the safety of structures such as schools, hotels, and office buildings. Not to dismiss these concerns, which are sometimes justified, but it is more often ordinary houses and apartment buildings that lead to infectious illness, not because of any dangerous materials used in their construction, but because of poverty, ignorance, or neglect.

Just imagine yourself in the following all too real scenario. A young couple are watching television one evening as their six-month-old child plays with a teddy bear. The proud parents frequently glance away from the television to look fondly at their beautiful, healthy child. All is well, until the child starts to bleed from one nostril. The father picks up the child and holds a tissue to its nostril, only to see the other nostril start to bleed as well. A short while later the bleeding stops, and the relieved parents decide that it was just one of those things. But when they check the child during the night, they find that the bleeding starts and stops for no apparent reason. By morning the child has begun to cough up blood, so the parents hurry to get dressed and race to the hospital. There the child is diagnosed with pulmonary hemorrhage and hemosiderosis (bleeding from the lungs), and dies within the next twenty-four hours.

Between 1993 and 1998 in Cleveland this scenario occurred not just once but thirty-seven times, with twelve fatalities. Thirty of the infants involved were African-Americans and lived on the poor east side of the Cleveland metropolitan area. Why thirty-seven healthy children should sicken and die without warning was a mystery. The "Cleveland outbreak" and similar incidents around the country began to be solved only when it was realized that the same black fungal mold was growing on moisture-ridden wallpaper or gypsum wallboard in all of the children's houses. This itself was puzzling at first, because there are molds almost everywhere and they commonly pose no danger to human health.

But the mold in the children's houses was not the ordinary kind. Called *Stachybotrys,* it can produce fungal spores containing poisonous satratoxins. The satratoxins can make anyone who inhales, ingests, or touches them sick, and in adults they produce such symptoms as skin and eye irritation, upper and lower respiratory distress, chronic fatigue, diarrhea, headaches, muscle aches, central-nervous-system disorders, and immune-system suppression. These stresses can in turn make adult sufferers susceptible to other infectious diseases. However, although somewhat controversial, current evidence indicates that satratoxin-producing *Stachybotrys* spores can be fatal to very young children because the rapid cell growth that occurs as the lungs develop gives the spores the best breeding ground they could ever hope to find. Infants who inhale these spores are at even greater risk if they live in a household with smokers, because tobacco smoke is one of a number of things that can trigger hemorrhages in their stressed-out lungs.

Fortunately, *Stachybotrys* is found in only two to five percent of American houses. Nevertheless, more than one hundred cases of hemosiderosis have been reported throughout the United States since 1993. The mold grows on the surfaces of wet materials that contain a high proportion of cellulose. In addition to wallpaper and gypsum wallboard, likely materials include fiberboard, wood, and cellulose ceiling tiles. The cellulose must be soaked with water for *Stachybotrys* to thrive, whether from leaky roofs, faulty plumbing, flooding, or some other source. It's worth noting that other fungi do well on moisture-soaked materials and that they too can cause health problems, if usually not fatal ones, for people with suppressed or compromised immune systems or chronic conditions such as asthma and other allergies.

The **Protective Response Strategy** for *Stachybotrys* and other fungi is therefore to prevent or quickly repair any residential water damage. If water damage has occurred, any cellulose materials should be removed and replaced before fungi have the chance to proliferate. The underlying cause of the moisture or water buildup then needs to be identified and corrected. The bottom line is that fungi require water to grow, so the drier things are the better.

Not everyone will have the ability or resources to make such repairs. Elderly people may not recognize that there is a problem before it is too late, for example. In poor neighborhoods, the negligence of slum landlords may be an exacerbating factor. We should never forget that our own health is connected to the health of everyone else in our communities. From a strictly selfish point of view, although very young children may die from inhaling *Stachybotrys* spores and thus present no health hazard to others, older children and adults may have their immune systems so weakened that they become prey to, and carriers of, other infectious diseases. Thus effective zoning and housing regulations need to be in place, and enforced, for the sake of a community as a whole. In addition, programs such as Habitat for Humanity can play an important role in raising the standard of community health by helping people to build better housing.

The need for good public-health regulations is also illustrated by the reconditioned mattresses we looked at briefly in the "Home Truths" tour in Chapter 6. I examined nine reconditioned mattresses for a segment on NBC's "Dateline," and found them to be riddled variously with filth, blood, urine, bed bugs, and many types of germs. The greatest health hazards were the fecal bacterium *Clostridium perfringens* and numerous fungi, all of which form spores that can filter through the bedding and become airborne. *Clostridium perfringens* spores, for example, could become an agent of gas gangrene in an ulceration or wound. For their part, fungal spores can trouble people with allergies, including asthma, those with chronic immunosuppressive or immunocompromising illnesses such as cancer or AIDS, pregnant women, and the very young and the elderly.

The insides of these mattresses were appalling, but how they got that way was equally shocking. As the "Dateline" segment chronicled, only nineteen states have regulations governing reconditioned mattresses, but even in those states the regulations are generally not enforced. New York State, for one, has not had inspectors to enforce existing laws on the sterilization and tagging of reconditioned mattresses for more than a dozen years. Manufacturers pay as little as five dollars apiece for discarded mattresses picked up on city streets or obtained from garbage dumps. Then they typically simply sew on a new cover, attach a tag claiming that the mattress has been "sterilized and reconditioned," and resell it. Or they cut off the existing cover, add some extra padding, resew the old cover on upside down, and then add a new cover with a "sterilized and reconditioned" tag. Needless to say, none of the reconditioned mattresses I examined showed any signs of having been sterilized or sanitized.

Reconditioned mattresses make up ten percent of the market. They are purchased for home use and, because of their lower cost, for some geriatric and pediatric facilities as well as some hotels and motels, insuring that members of the most vulnerable population groups will come into contact with them. The appropriate **Protective Response Strategy** is, once again, to invest in new mattresses, if household finances allow, or to seal reconditioned mattresses inside plastic or heavily woven, impervious coverings, which will create a barrier against infectious germs and allergens. Better regulation of the reconditioned-mattress industry would also be desirable, of course.

Too Clean for Our Own Good?

The potential that fungi and other germs have for triggering asthma and allergy attacks raises the general question of how to deal with these chronic, often debilitating health problems. In the United States one in five adults suffers from asthma or other allergies, and all told more than fifty million people have them. One hundred thirty million school days a year are lost in America because of asthma and allergy attacks. And the problem is getting worse: Asthma rates in the United States have increased 158 percent over the past two decades.

Allergies and asthma (asthma is really a form of allergy) are hyperimmune responses to innocuous or relatively innocuous foreign substances. These conditions can easily be aggravated by allergens that come from pets, cockroaches, fungal molds and spores, other germs, pollen, and dust mites. Up to eighty percent of household dust is made up of the common dust mites *Dermatophagoides*. Mattresses and other bedding provide important breeding grounds for these mites, especially in hot, humid weather. One study found that dust from spring mattresses contained more mites than dust from foam mattresses. Even waterbeds can have significant levels of dust mites. Another major study found that one hundred percent of household dust samples also contain fungal mold spores, such as *Penicillium, Aspergillus, Alternaria, Cladosporium,* and others, that can trigger allergic reactions. Finally, pet allergens can be present even in houses without pets, because pet owners trail them everywhere they go. Taken together, these microscopic airborne particles can be devastating to people with allergies such as asthma.

All this suggests that people with asthma and other allergies should keep their houses and themselves as clean as possible. But in the last few years a controversy has arisen as to whether cleanliness in early life might actually help cause these ailments. Some researchers contend, on the basis of very limited evidence, that allergy and asthma are "the price of hygiene." They argue that asthma rates are higher in children in the United States than in Third World countries, for example, because obsessive cleanliness has cut them off from "natural" exposure to too many infectious germs, thus preventing their immune systems from developing properly. One study found that Italian air force cadets who had antibodies from exposure to three food-borne germs (*Toxoplasma, H. pylori,* and Hepatitis A) had fewer allergies than cadets who lacked the antibodies. There is an illogical inference made here regarding cleanliness and allergies. No amount of handwashing will prevent ingestion of toxoplasmosis germs already present in raw meat. Such germs are transmitted because of improper cooking, *not* improper handwashing or personal hygiene! Likewise Helicobacter pylori can be transmitted by oral contact or by water. No amount of handwashing will prevent this. A study in Arizona similarly claims that children in day-care centers develop less asthma, because daily exposure to germs from other children strengthens their immune systems. The scientist who conducted this study makes a big point of the fact that childhood asthma rates seem to be lower in the Third World than in the United States.

Carried to its conclusion, the logic of this argument is that the oftener young children get sick, the better for them in the long run. There's a kernel of truth in that. As I have already explained, children need regular exposure to environmental germs and to the normal flora of other people in order to develop healthy immune systems. But let's not throw the baby out with the bathwater on this issue. Otherwise we will put ourselves—or, to be more accurate, our children—in the position of ignoring the real threats that childhood infections can represent. Remember rotavirus, which we discussed in Chapter 6. When it spreads through a day-care center, it can make some children so sick that they require hospitalization, and in the worst cases it can kill them. In the Third World, rotavirus regularly kills countless young children.

One reason that childhood asthma rates seem to be lower in the Third World is very likely the fact that several million Third World children a year die at very early ages from diarrheal illness alone. Add the toll of other infectious diseases, and it clearly becomes a question whether many Third World children simply don't live long enough to be recorded as having asthma or other allergies. It could well be that asthma rates would rise in the Third World if infant and childhood mortality decreased. This would be a cause for celebration, not misplaced nostalgia for the dirty ways of the past.

The "too clean for our own good" argument seems to me to rest on a fallacy. Commonsense cleanliness does not, and could not, isolate children from exposure to environmental and human germs. The germs are far too plentiful for that ever to happen. My parents kept a very clean household, and they taught me to practice good personal hygiene. That didn't prevent me from being exposed to normal childhood illnesses and coming down with my fair share of them in our old neighborhood in Brooklyn. And like the majority of people, I did not develop asthma or other allergies.

One set of circumstances more than any other is associated with asthma in the United States, and that is poverty. Asthma rates are by far the highest among children in poor, dirty neighborhoods with high rates of infectious disease and housing that is all too frequently infested with cockroaches and other vermin. This fact alone should be enough to make the "don't be too clean" brigade pause and reconsider their advice. Fortunately, conditions even in the poorest urban areas in America tend at least to be better than those in the Third World, because of good sanitation, decent nutrition, and greater access to health care.

If we are looking for an explanation for the 158 percent increase in asthma in the United States over the past two decades, we might reasonably begin by noticing that the same period has seen a large increase in the number of people living in poverty, thanks to big increases in immigration. I speak to this issue, once again, as a medical scientist and as the child of an Italian-American immigrant family. If we are going to continue to bring hundreds of thousands of poor people into the country every year, we have to face the health aspects of this policy. The new immigrants are becoming our neighbors. If too many of our neighbors live in unhealthy conditions, sooner or later that will affect us all.

Of course, asthma and other allergies, like infectious diseases in general, affect people in every socioeconomic group. But asthma, among other health problems, is far more common at the bottom rungs of the ladder than at the top, and this relationship needs to be examined from every possible angle. With respect to allergies in general, medical science still has a great deal to learn about how these hyperimmune responses are triggered and why some people, rich and poor alike, are more susceptible to them than others.

Until cures for asthma and other allergies are found, household **Protective Response Strategies** can make attacks less likely and reduce their severity.

- **To combat mattress and bedding allergens, follow the Protective Response Strategies on page 165.**

- **Vacuum clean floors, drapes, and upholstery regularly to reduce mites and other allergens. Be sure your vacuum cleaner works properly and is fitted with air filters that prevent allergens and dirt from being recirculated into the household air.**

Central vacs are particularly good, because they collect debris and vent it at a remote area of the house.

- Because they act as collection areas for allergens, wall-to-wall carpeting, heavy upholstery, and thick drapery should be avoided, if possible. Some interior designers contend that "less is more"; for allergy sufferers, less is better.

- Carpets in particular can become heavily infested with mites. The chemical benzyl benzoate will kill mites, but their dead carcasses can still be allergenic and will need to be vacuumed up afterward.

- Washing and drying at temperatures greater than 55 degrees centigrade (about 131 degrees Fahrenheit; the "hot" cycle of the washer) will also kill mites, as will dry cleaning.

- A good air-filtration device can be a great help in controlling dust mites and fungal spores, especially in sleeping areas, and humidifiers can maintain the integrity of our mucosal systems so that they can operate properly.

- Respond quickly to any sign of cockroaches or other vermin (in the case of rodents, remember the Protective Response Strategy on page 155) with appropriate pesticides and traps.

- Pets pose a special problem. Some pet lovers with asthma or other allergies may gradually develop a tolerance to individual animals, or be able to manage their symptoms with allergy medicines. If it becomes necessary to give up a pet, it will take four to six months for pet allergens to fall to normal background levels.

9

On the Wing and on Eight Little Legs

The Canary in the Mine

In the old days miners sometimes carried a caged canary underground with them. More sensitive than a person to atmospheric changes, the canary functioned as an early-warning system. If the little bird showed signs of distress or died, the miners knew they had to flee as quickly as possible before poison gas or a lack of oxygen struck them down, too.

Birds can also serve as an early-warning sign of environmental danger above ground. Alfred Hitchcock played with this notion to terrifying effect in his movie *The Birds,* in which flocks of birds attack the inhabitants of a small town on the coast of Northern California. Struggling to understand what is happening, and to survive the birds' attacks, the characters in the movie conclude that human pollution of the environment must be the ultimate source of the trouble. In fact, birds and other small animals often do give warning of an ecological disturbance, not because it turns them into crazed killers, but because they are among its first, most noticeable victims.

This happened in New York City in the summer and fall of 1999. From midsummer on, local health officials began to notice an increase in the deaths of local birds, especially crows. At the Bronx Zoo, a cormorant, two Chilean flamingos, and an Asian pheasant all died mysteriously. When the bodies of the zoo birds and a crow were examined, evidence of encephalitis and meningitis was found, but they tested negative for all the viruses that usually cause these conditions.

Meanwhile, on August 23, 1999, an infectious-disease physician in the New York City borough of Queens, Dr. Debbie Asnis of Flushing Hospital, reported two patients with unexplained encephalitis. The New York City Department of Health investigated and found a cluster of six unexplained cases in the Whitestone/Flushing area of northern Queens, on the shore of Long Island Sound. Relatively healthy adults older than fifty-eight years of age were falling ill with flu-like symptoms of fever, headache and muscle ache, nausea, vomiting, and diarrhea. The symptoms rapidly progressed to include altered mental states of confusion, stupor, and even coma.

On August 31, I received a New York City Department of Health fax and E-mail alert about the puzzling bird deaths and cases of encephalitis in human beings at my office in the Department of Clinical Microbiology at NYU Medical Center. The alert asked everyone involved with infectious medicine to be on the lookout for people with similar symptoms and to collect specimens for analysis in laboratories at the New York State Department of Health. A week later, on September 6, a follow-up alert reported the deaths of more birds and thirty-six cases of suspicious encephalitis in people. The encephalitis patients came from throughout the New York City region, but most of them lived in the Whitestone/Flushing area. Two elderly patients had died, and tests at the CDC identified the cause of their illnesses as Saint Louis encephalitis virus (SLE). This was a shock, because SLE had never before been found in New York City.

One of the flavivirus group of viruses, SLE gets its name because it was first identified in Saint Louis in 1932. During the summer months, SLE periodically causes illness throughout the southern half of the United States. Before 1999 its northernmost reported occurrence was in New Jersey. Apparently it had moved a little farther up the coast and crossed the Hudson and East rivers into New York City and, more specifically, the borough of Queens. Or had it? As more specimens from the encephalitis outbreak were examined, it emerged that only some of them tested positive for SLE. Others tested negative both for SLE and for other encephalitis-causing viruses known to be present in the United States.

It can often be difficult to pin down the specific cause of viral encephalitis or meningitis. A wide array of viruses can trigger these illnesses, which can also be caused by bacteria. The usual test involves examining spinal fluid drawn from a spinal tap to look for traces of germs affecting the central nervous system. Bacteria can be identified, or ruled out, fairly straightforwardly. But if a virus is involved and investigators don't have at least a rough idea of what they're looking for, the test can easily come up empty. Investigators can also be fooled, even if they're looking for something in the right group of viruses, because closely related germs can be cross-reactive. Two substances are cross-reactive when they elicit reactive responses from the same antibodies in the immune system. This phenomenon may involve the body's own tissues. Group A streptococcus triggers responses from an antibody that will also react to and bind with cells in heart tissue, for example, and this can lead to rheumatic fever that damages heart muscles and valves.

Finally, on September 23, a month after the first human cases of unexplained viral encephalitis were reported in Queens, testing at the CDC revealed that the culprit virus was not SLE but another flavivirus, West Nile virus (WNV), so called because it was first identified in 1937 in a sick woman in the West Nile district of Uganda. This was even more of a shock, because West Nile virus had never before been found in the Western Hemisphere, let alone New York City!

West Nile on the Hudson

How West Nile virus reached the shores of the Hudson River depends crucially on its mode of transmission. Viruses like SLE and WNV usually require an insect vector to move from host to host. These and many other viruses are commonly spread by mosquitoes and ticks, but mosquitoes are the primary vector for both SLE and WNV. In the wild their main hosts are birds, and a bird-mosquito cycle is what keeps the viruses going. Many other ani-mals, including bats, cats, dogs, cattle, camels, horses, mice, frogs, and people, can become incidental hosts when they are bitten by virus-bearing mosquitoes.

It would be hard to design a more efficient vector for an infec-tious germ than the mosquito—or, to be more accurate, the female mosquito. Male mosquitoes feed only on flower nectar, but adult females also feed on blood and usually require a blood meal to pro-duce a batch of eggs. In taking that blood meal, female mosquitoes offer any germs they carry direct entry into a host's bloodstream. Individual mosquitoes may bite many birds, animals, and people, picking up and transmitting germs as they go, before laying their eggs and completing their life cycles. When the larvae emerge from the eggs, they are equipped with the full complement of their mothers' germs and can extend the chain of infection.

The usual life span of a mosquito is one to two months, but this can easily double during periods of drought. The mosquito holds on longer during dry spells, because its larvae can only live in water. If there is no water available, a mosquito may lay its eggs on ground that was previously flooded; these eggs can remain viable for up to five years and hatch when reflooding occurs. Mos-quitoes also need relatively warm weather, but if a frost and sus-tained cold arrive before they have laid their eggs, they can go into a state of hibernation and last out the winter, so long as it is not too severe. As a result, mosquitoes have been able to shadow human settlement and establish themselves in all but the most extreme habitats. In the process, they have carried viruses and other infectious germs from one corner of the globe to another.

In their original habitats, viruses like WNV and SLE do not kill many birds or other animals. Over countless generations, these viruses and their main bird hosts accommodated themselves to one another in ways we've explored in earlier chapters. A para-site's goal is not to kill its host, after all, but to survive and repro-duce in it for as long as possible. As parasite and host coexist, or perhaps the better word is "coevolve," survival of the fittest will gradually favor less lethal strains that keep the host alive to sup-port the parasite. Likewise, exposure to an infectious germ stimu-lates a creature's immune system to produce protective antibodies.

The longer an infectious germ exists in an environment, therefore, the likelier potential hosts are to develop a tolerance for it. The flip side of the coin is that in a new environment, an invasive virus can be deadly to creatures that have never encountered it before.

This has been the pattern for WNV so far. After the virus's discovery in Uganda in 1937, it cropped up in mosquitoes, birds, people, and other animals in Egypt and Israel in the 1950s, and soon thereafter in other parts of the Middle East and in western Asia, Europe, and even Australia. Serious human epidemics broke out in Israel during 1950-54 and in 1957, and in France's Rhone River delta in 1962. One of the largest outbreaks occurred in 1996 near Bucharest, Romania, where five hundred people fell ill and fifty died. Most people who are infected with WNV either develop no symptoms or suffer only mild fevers. But as with the elderly victims who have died in the New York City area, the members of a population with weaker immune systems are always at greater risk.

Until quite recently WNV had not produced any unusual bird deaths in Africa, the Middle East, Europe, or western Asia, as it has in the United States. This may be because birds migrating to and from Africa introduced the virus gradually to bird populations in these areas, lessening the impact that it might have had if it hit all at once. Another possibility is that the WNV that has come to America may be a new and more potent strain. In 1998 a large number of birds in Israel died in an epidemic, or epizootic as animal epidemics are called, of WNV. When the CDC finally identified the cause of 1999's mysterious animal and human encephalitis cases in and around New York City, it found that the WNV involved was identical with the strain isolated in the 1998 avian outbreak in Israel.

This observation has led some researchers and public-health officials to conjecture that a person or animal infected in Israel carried the virus to the New York area. But I suspect another possibility may be equally likely. To date, wherever WNV has been found in the United States, an Asian mosquito called *Aedes (Oclerotatus) japonicus* has also been found. The common American mosquito is called *Culex pipiens*. After the outbreak in New York City, U.S. army researchers determined that the *Aedes* mosquito transmits WNV four to five times more efficiently than *Culex* does. This suggests that the most severe West Nile fevers stemmed from bites by

Aedes rather than *Culex* mosquitoes. In addition, the *Aedes* mosquito bites throughout the day, whereas *Culex* is active only from dusk to early morning.

Aedes mosquitoes have only rarely been found in the northeastern United States before 1999, but on October 29, 1998, a year before the first cases of WNV in the United States were identified, the New York City Department of Health reported that an area resident had contracted a case of eastern equine encephalitis (EEE) from the bite of an *Aedes* mosquito. (Despite its name, EEE virus is mainly found in birds in the southeastern United States; horses are only incidental hosts.) With these facts in mind, let me tell you a scary animal fable, an imaginary but very possible scenario.

Five Mosquitoes Cross the Sea

Tel Aviv, Israel, June 6, 1998: One of El Al's daily flights to New York City readies for departure, and five mosquitoes go along for the ride. Mosquitoes and other insects travel on airplanes all the time. One study found that twelve of sixty-seven commercial airliners arriving in London from tropical countries had live mosquitoes on board. *Anopheles* mosquitoes, the vector of malaria, are such frequent (airline) fliers that they create the phenomenon of "airport malaria," when they bite people who live in the vicinity of their destination airports. Among western European countries, France has the highest incidence of airport malaria, about thirty cases over the last thirty-six years, some of them fatal.

The five mosquitoes on today's El Al flight are not *Anopheles* mosquitoes, however, but *Aedes*. Sisters from the same hatch, they all carry a strain of WNV that has just killed many Israeli birds. Three of them fly into the passenger cabin, attracted by the pheromones, body heat, and carbon-dioxide exhalations of the passengers and crew, and two enter the cargo hold, where they buzz around cases of hothouse flowers that are bound from a kibbutz to a wholesale florist in Manhattan.

During the flight one mosquito in the passenger cabin bites the ankle of a fifty-eight-year-old product manager, who is on his way home to College Point, Queens, from a business trip, and then settles into the man's trouser cuff. The second mosquito flits about between two passengers and finally decides to bite a young pregnant woman, a bookkeeper from Flushing, Queens, who has been on a tour of the Holy Land with her husband and a group of fellow parishioners from their church. Pregnant women are warmer than other people because the flow of blood to the skin increases during pregnancy, and the mosquito is attracted to this greater warmth. After biting her, the mosquito goes to sleep in the thick pile of her sweater. Occasionally both mosquitoes reemerge for another bite, although one bite is usually enough to transmit an infection. The third mosquito is less active and does not bite anyone.

When the flight reaches New York it lands at JFK Airport in southern Queens. The mosquito in the product manager's trouser cuff takes a cab ride with him to his home in College Point, Queens, on the edge of Long Island Sound and within hailing distance of the Whitestone Bridge. At this point, the mosquito takes a final meal from her unwitting blood donor and flies off to lay her eggs in an inlet of the sound. The mosquito on the bookkeeper's sweater rides with the woman and her husband to their home in Flushing. There the mosquito flies away and after briefly wandering around bites a mechanic at an auto-repair and tire shop. She lays her eggs behind the shop in a puddle of warm water inside a discarded tire. The third mosquito remains in the passenger cabin and is killed when the plane is routinely sprayed with insecticide.

Meanwhile the two mosquitoes in the cargo hold have been impatient for freedom. As soon as the cargo-bay doors are opened, they zoom out. In the Gateway National Recreation Area adjacent to the airport in Jamaica Bay, they find birds aplenty to feed on and acres of marsh where they can lay their eggs. When the eggs of all four mosquitoes hatch, the larvae that emerge will bear WNV. If they survive to adulthood, the females among them will be able to transmit the virus to every living creature they feed on, from birds and bats to raccoons and horses to cats, dogs, and people, among many potential others. By the same token, any local *Culex* mosquitoes that bite infected animals and people will also acquire WNV and become able to transmit it, although they will do so less efficiently than infected *Aedes* mosquitoes. (Human blood may stunt egg or viral development in some mosquito species.)

Within a few days of the flight, both the fifty-eight-year-old product manager and the young bookkeeper feel like they're coming down with the flu. The man's initial symptoms include a mild fever, an unrelenting headache, and a stiff neck. When the symptoms worsen, the man's wife insists on taking him to the emergency room of a local hospital. The emergency room staff do a spinal tap, find no traces of bacteria, and make a tentative, default diagnosis of aseptic meningitis, "aseptic" indicating the apparent absence of any pathogenic germs. It would be pointless to look for evidence of a virus without some clue as to what kind of virus it might be; there are, again, so many possible viral sources of encephalitis and meningitis that it would be like looking for a needle in a haystack.

For her part, the young bookkeeper gets only slightly ill. Convinced that she caught a "flu bug" from a fellow passenger on her flight, she stays home from work for a few days to rest, grateful that she has an understanding boss. As she recovers, she spends part of each day in her backyard, reading a Stephen King novel and napping off and on. On the first day she is bitten by a number of indigenous *Culex* mosquitoes, so before she goes back out the next day she makes sure to put on some insect repellent. The local mosquito population is bigger than normal this year, because heavy spring rains have favored their reproduction. The *Culex* mosquitoes that fed on the bookkeeper before she used repellent have picked up the WNV in her blood, and they can now pass it on to the creatures they bite and to their progeny.

Over the next year, WNV gains a solid foothold in the Flushing/Whitestone/College Point area and spreads farther as people and birds travel to and from other parts of the greater metropolitan area. The weather plays a significant part in the process. The winter of 1998-99 is a mild one, allowing more virus-bearing mosquitoes to survive in hibernation until the spring. Although the summer of 1999 turns out to be relatively dry, cutting down on the number of mosquito hatches, the mosquitoes extend their life spans so that they can lay their eggs in favorable conditions and insure that new virus-bearing mosquitoes are born. The virus also survives in the infected birds, animals, and people, probably the great majority of those infected in each group, that it does not kill.

Again, most people infected with WNV suffer at worst a mild fever and other flu-like symptoms. But those who survive infection become reservoirs of the virus and can become links in new chains of infection whenever they are bitten by mosquitoes. In addition, scientists at the U.S. Geological Survey's National Wildlife Health Center have shown that WNV can be transmitted directly from bird to bird by normal pecking and other behaviors, at least in a laboratory setting. That finding suggests that transmission from birds to human beings may also be possible.

Something like my fable of five *Aedes* mosquitoes crossing the sea—or like the scenario of an infected person or animal traveling or being brought from Israel to New York City—almost certainly occurred in 1998. Given that the WNV found in New York was identical with the strain responsible for the 1998 outbreak in Israel, it probably could not have arrived sooner. Certainly it would have taken several months to a year for the virus to extend its reach as far as it did by September 1999. After the alert was sounded in New York City, WNV was quickly identified in seven states: Connecticut, Maryland, Massachusetts, New Hampshire, New Jersey, New York, and Rhode Island. In each of these states the rare *Aedes* mosquito was also found. Finally the human case of EEE reported by the New York City Department of Health in October 1998, and linked to the *Aedes* mosquito, provides a firm date for that foreign species' unusual presence in the New York area.

WNV now seems to be in the United States to stay. By the time the first sustained frost ended the 1999 mosquito season, there had been sixty-two cases of WNV-induced encephalitis in the New York metropolitan area, with seven deaths. Although there were no human fatalities in the other affected states, the toll on wildlife was considerable. It is clear that hundreds of birds of varied species, such as crows, herons, hawks, and blue jays, died from WNV in 1999, along with an uncertain number of small mammals such as bats, raccoons, and squirrels. Many birds and animals must have died in the wild without their deaths ever being noticed or recorded.

The numbers would surely have been greater if municipalities and public-health departments had not engaged in an aggressive public Protective Response Strategy. The rapid initial response spearheaded by New York City's Mayor Rudolph Giuliani was exemplary. This combined surveillance for new cases of illness with a campaign to contain and reduce the mosquito population by eliminating pools of standing water, applying larvicides, and spraying with malathion and other insecticides. Efforts to inform the public of the advisability of personal **Protective Response Strategies,** such as avoiding mosquito-infested areas, wearing long-sleeved shirts and long trousers, using mosquito repellent, putting screens in windows, and changing the water in birdbaths regularly probably helped as well.

The public **Protective Response Strategies** were suspended over the winter of 1999–2000 and reinstituted with the return of warmer, mosquito-friendly weather. The results were both encouraging and discouraging. On the one hand, the number of people made ill by WNV in New York City and the surrounding region dropped considerably. There were only fourteen cases of West Nile fever, with one death, in 2000, as opposed to sixty-two, with seven deaths, in 1999. Interestingly, the majority of the cases in 2000 were no longer in the Flushing/Whitestone area near Long Island Sound, where the most aggressive spraying was done, but on Staten Island, which is more or less at the opposite end of the city. On the other hand, WNV extended its geographic reach from seven states to twelve—Connecticut, Delaware, Maryland, Massachusetts, New Hampshire, New Jersey, New York, North Carolina, Pennsylvania, Rhode Island, Vermont, and Virginia—and the District of Columbia. The virus killed 4,139 birds of 76 different species, from gulls, herons, and cranes through crows, blue jays, and robins to hawks, doves, and bald eagles. That is, it killed that many that we know of; many others presumably perished unnoticed in the wild, along with numerous small mammals.

Over the same period the strain of WNV found in the United States caused more and more trouble in Israel, where it seems to have originated. In 1999, Israeli agricultural authorities had to order the destruction of eight thousand geese when West Nile fever was discovered in commercial flocks. And in August and September 2000, Israel's Ministry of Health reported 151 cases of West Nile fever among elderly Israelis, with twelve deaths.

Although these data indicate that New York City–area health

authorities were right to implement an aggressive program of spraying with insecticides in 1999 and 2000, mosquito control can be a two-edged sword. Once we release insecticides and pesticides into the environment, we have no further control over them. A substance intended to kill one creature may harm others we don't want to hurt. As it happened, the WNV spraying coincided with the loss of ninety percent of the lobster population, about eleven million lobsters, in the waters of Long Island Sound. Other crustaceans, such as crabs and starfish, also died in great numbers. It is likely that the next generation of crustaceans was poisoned as well by insecticides leaching into the sound. For all their apparent differences, lobsters, crabs, starfish, and mosquitoes are all arthropods, and so are vulnerable to the same toxins.

It would be good to have more environmentally friendly **Protective Response Strategies** to employ. Perhaps we can devise lures and traps, for example, that capitalize on mosquitoes' attraction to human pheromones, carbon dioxide exhalation, and body temperature. Or we could stock bodies of water where mosquitoes breed with fish that eat mosquito larvae. Making such subtle approaches work on a large scale will obviously be a daunting technological and scientific challenge. But as an exploding human population continues to make large-scale changes in the environment, thereby opening up new avenues of infection for germs and their vectors, such as plane travel, it becomes ever more important for us to find ways of working with nature rather than running roughshod over it.

Of Mosquitoes and Men

No one can yet tell what the ultimate extent of WNV will be. But infectious germs using mosquitoes to access their human prey is nothing new. The bloodsucking mosquito has been a successful human parasite in its own right for thousands of years, and during that time many germs have found it to be the perfect vector for their needs, especially when they are aided and abetted by human behavior and human changes in the environment. There is no better example of this than the history of malaria.

Malaria is caused by a microscopic protozoan parasite known

as *Plasmodium.* Transmitted by the bite of *Anopheles* mosquitoes, *Plasmodium* moves via the bloodstream to the liver, where it breeds for a couple of weeks. Once the *Plasmodium* has multiplied sufficiently, it returns to the bloodstream and begins to eat away at and reproduce inside red blood cells, causing waves of violent chills and high fever. As the disease progresses, the liver and spleen become enlarged, and death is often the ultimate result.

Plasmodium and *Anopheles* may have been intimately connected for aeons. But their partnership against human beings took off in a big way only five to ten thousand years ago, when agriculture was introduced to Africa. Slash-and-burn agricultural techniques and animal husbandry created ideal breeding grounds for mosquitoes, turning forests and savannas into a landscape of shallow ponds and puddles of standing water. Ever since then *Anopheles* and *Plasmodium* have followed people into all but the coldest climates. Malaria was endemic in much of the ancient world, from Africa and the Mediterranean through the Middle East to India and China.

In February 2001, scientists at the University of Manchester in Great Britain reported that they had recovered malarial DNA from the skeletons of people buried in mass graves near Rome around A.D. 450. Contemporary accounts of the time spoke of terrible fevers killing large numbers of people. The new finding explains these fevers and also solves a puzzle that has long perplexed historians. In A.D. 452, Attila the Hun was leading an army toward Rome, which lay virtually defenseless against him. Yet he suddenly turned his army around and left the city untouched. It now seems that Attila wisely feared the fierce epidemic that was then raging through the heartland of the Roman empire and that would contribute significantly to its ultimate decline and fall.

Malaria first reached the New World of the Americas during the 1500s in the blood of Spanish conquistadores. Then it arrived in wave after wave in the blood of West African slaves, as well as that of European settlers and traders. The result was that malaria became endemic not only in the Caribbean basin but in much of what would become the United States, sometimes raging as far north as New England. Inland, malaria advanced with the frontier as settlers cleared forests for farming.

The connection between malaria and the slave trade has a fascinating genetic component. Malaria would not have become established in the Western Hemisphere as firmly or as quickly as it did if West Africans, who had a high incidence of the disease, had not been transported there as slaves. Yet West African people's desirability as slaves stemmed in part from the fact that many of them were resistant to malaria because of a genetic trait, called sickle cell, that a substantial percentage of African-Americans continue to bear. The sickle-cell trait involves a mutated gene for hemoglobin molecules, which carry oxygen inside red blood cells for transfer to body tissues. When the mutated hemoglobin transfers its oxygen, it tends to rupture the red blood cells, giving them a sickle-like appearance. The sickling depletes potassium, which the *Plasmodium* parasite needs to reproduce inside the red blood cells. This gives the immune system time to mount a counterattack against the parasite, and so confers some protection against the worst ravages of malaria.

The mutated gene spread quickly after the introduction of agriculture put malaria into overdrive, because people who had it lived longer. Its presence in the West African population created another problem, however. Like other creatures that reproduce sexually, human beings receive copies of two sets of genes, one from the father and one from the mother. A single copy of the sickle-cell gene confers resistance to malaria. But two copies confer the potentially deadly disease of sickle-cell anemia. So although it is a piece of good fortune for people born in malarial areas if one parent has the mutated gene, it is very bad luck indeed if both parents have it to pass along in their DNA.

Sickle-cell and a number of similar genetic adaptations to *Plasmodium* lessened the impact of malaria in some population groups. But the disease remained a potent killer even in the best of genetic circumstances. The noted twentieth-century Australian immunologist McFarlane Burnet held that, of all infectious diseases, "there is no doubt that malaria has caused the greatest harm to the greatest number." Although AIDS and tuberculosis destroy more lives today, over the course of human history perhaps only smallpox can claim equal killing status with malaria.

The toll exacted by malaria made figuring out the process of the disease a chief goal of early microbiology. The pieces of the puzzle finally came together at the end of the nineteenth century, thanks to a Frenchman, Alphonse Laveran; an Englishman, Ronald Ross; and an Italian, Giovanni Grassi. In 1880, Laveran discovered the *Plasmodium* parasite while working as an army surgeon in Algeria. In 1897, Ross, working as an army surgeon in India, found *Plasmodium* in the stomachs of *Anopheles* mosquitoes. A year later, Grassi also identified *Anopheles* as the vector of malaria; he then figured out the entire chain of infection and the whole of *Plasmodium*'s life cycle. These were Nobel Prize–worthy achievements, but although all three scientists made at least equal discoveries, only Ross and Laveran received the prize.

Identifying mosquitoes as the vector of malaria infection opened up a promising new chapter in humanity's struggle with tropical diseases. Subsequent research soon established that mosquitoes are also the primary vector for dengue—or breakbone fever, as it is memorably called for the excruciating pain that it causes—yellow fever, and other ailments. Together these findings led to efforts to control the diseases by controlling, or even eradicating, the populations of various species of mosquitoes. The first target was yellow fever.

European settlement and the slave trade brought yellow fever to the Americas along with malaria, but the two diseases are transmitted from one host to another by different mosquito species. Whereas *Anopheles* is the vector for malaria, *Aedes aegypti* is the vector for yellow fever, which causes internal hemorrhaging, jaundice, coma, and ultimately often death. Jaundice's yellowing of the skin gives the disease its name. From the early colonial period to the beginning of the twentieth century, yellow fever periodically swept through port cities along America's east coast as far north as Boston, as well as up and down the Mississippi River valley. In 1793 a yellow fever epidemic killed four thousand people in Philadelphia out of a population that then numbered about forty thousand. The epidemic spread throughout the northeastern part of the country and then moved south, waxing and waning over the course of the next twelve years. When it reached New York, in 1793, Governor George Clinton (no relation to President Bill Clinton) was forced to issue a proclamation quarantining all yellow fever victims. Because the role of mosquitoes in the disease was not yet understood, it was feared that yellow fever could be transmitted directly from person to person.

The prevalence during this period of yellow fever, among other infectious diseases such as malaria, helps explain why only sixteen percent of the people born in Philadelphia during the 1790s survived past age thirty-six. Between 1800 and 1850 there were fifteen epidemics in Savannah, twenty-two in Charleston, and thirty-three in New Orleans. After an epidemic in New York City in 1819, the Public Workhouse and House of Correction on Bell Vue Place, which was originally built in 1736 and which had a wing for the sick, expanded into a new fever hospital that became the core of what is now Bellevue Hospital on the city's lower east side. In 1878 an epidemic stretched for seven hundred miles from New Orleans to Saint Louis.

As early as 1807 some observers were implicating mosquitoes in the disease process, whereas others insisted that miasmas were the source of the illness. As we've seen, it would take another ninety years for medical science to reach the point where it could prove the mosquito connection. Until then the only protective response people had was to quarantine the victims until they died or recovered. Ships with yellow fever on board were required to fly a yellow pennant, or "jack," and yellow jack became another name for the disease.

Once mosquitoes were conclusively linked to yellow fever, however, eradication schemes quickly followed. The U.S. military led the way, spurred on by the appalling number of soldiers who died of yellow fever during the Spanish-American War of 1898. In 1901, the military mounted an all-out assault against yellow fever in Havana, which was then occupied by U.S. troops. In addition to quarantining people with yellow fever, the campaign included draining pools of standing water, spreading kerosene on ponds to smother mosquito eggs, and covering every well, tank, and water barrel in the city. Within three months, yellow fever was eliminated from Havana.

In 1904, the United States took over the building of the Panama Canal and began to apply the same tactics to protect workers. France had abandoned the project a few years earlier, in no small part because of yellow fever deaths. Six hundred or more workers might die of the disease in a single month, and many people felt that it was futile to try to combat mosquitoes in the Panamanian jungle. General G. W. Davis, the governor of the Panama Canal Zone, maintained that "spending a dollar on sanitation is as good as throwing it into the Bay." But mosquito eradication turned out to be so effective that September 1906 saw the last yellow fever death among Canal Zone workers. When the canal was finished in 1913, it was a healthier environment, in regard to mosquito-borne diseases, than most of the southern United States.

Encouraged by these successes, American medical authorities launched a series of campaigns against mosquitoes. The Rockefeller Foundation and the United States Public Health Service started this effort in Arkansas, using copper arsenic to kill mosquito larvae. By 1927, the United States, parts of which were once almost as malarial as equatorial West Africa, was virtually malaria-free. Yellow fever, and the *Aedes aegypti* mosquito that carries it, were also dealt a crushing blow. This may explain why *Aedes* mosquitoes, like the *Aedes japonicus* which spreads WNV, have been rare in the United States until recent decades, when commercial airliners began to ferry mosquitoes around the world. For example, the Asian tiger mosquito, *Aedes albopictus,* which carries a number of disease-causing germs, appeared in Houston in 1985, and its numbers have been increasing steadily ever since.

Other countries soon emulated the United States, in some cases with the assistance of the Rockefeller Foundation, which in 1913 initiated a worldwide program for control of yellow fever, malaria, and parasitic hookworms. (John D. Rockefeller had become interested in funding medical research in 1901, when his first grandson died of scarlet fever, and he established both the Rockefeller Institute, now Rockefeller University, and Memorial Sloan-Kettering Cancer Center.) Italy, which had two million malaria cases a year around 1918, copied the American model with great success in the 1920s and 1930s. As the century advanced, public-health officials began to hope that malaria might be brought under control everywhere, especially with the help of new, more powerful pesticides like DDT. In the late 1950s, the U.S. Congress appropriated funds for a worldwide anti-malaria campaign intended to eliminate the disease by 1963.

It was not to be. We learned to our horror that DDT and similar pesticides would not only affect mosquitoes, but would threaten the entire environment and our own food chain. In 1963, perhaps thinking it had done enough, Congress stopped appropriating funds to combat malaria outside the United States, giving the disease new impetus just when the number of cases had been brought to their lowest point. Most troubling of all, the *Anopheles* mosquito became resistant to DDT, and subsequently to many other pesticides, and the *Plasmodium* parasite became resistant to quinine, chloroquine, and other drugs intended to moderate its effects. These two partners in malaria have rolled with every punch that human beings have tried to throw, and more often than not they have counterpunched with devastating power. The net consequence of the twentieth century's efforts to control *Anopheles* and *Plasmodium* is that there is now three times as much malaria in the world as there was in the late 1950s and early 1960s. Although malaria remains relatively rare in the United States and other developed countries, at any one time about 500 million people around the world suffer from the disease. More than two million people die of it every year, nearly one million of them children in sub-Saharan Africa. Every thirty seconds a child dies of malaria.

Research to find a vaccine has so far not yielded much in the way of effective results. But this may well change as new molecular and genetic techniques give medical scientists the ability to look inside the cells of both *Plasmodium* and human beings and observe their interaction in precise detail. At NYU Medical Center, for example, Drs. Ruth and Victor Nussensweig are leading an effort to develop an experimental malaria vaccine that will marshal the immune system's cytotoxic T cells, antibodies, interferon, and other cytokines against the distinct developmental stages of the *Plasmodium* parasite.

A vaccine is sorely needed, because the threat of malaria, as well as other mosquito-borne diseases, will likely increase in the years to come. A globalizing economy with a billion human air travelers a year (and sizable numbers of hitchhiking mosquitoes), unbridled population growth, and massive levels of migration are extending the process of environmental disruption and degradation that humanity initiated with the first slash-and-burn agricultural techniques ten thousand years ago. The same techniques are still being used in Africa and Brazil, although human beings now have machines to accelerate the slashing, burning, and timbering that turn vast stretches of rain forest into unproductive, thinly soiled farmland and prime mosquito breeding grounds. It is no coincidence that malaria rates have skyrocketed in Brazil, for example.

Global Warming

The most serious factor affecting the future of mosquitoes and the infectious germs they carry is global warming, which threatens to turn temperate regions, including much of the United States, into subtropical zones. During the 1990s, the hottest decade in a thousand years, the Earth warmed one full degree Fahrenheit. Some experts predict that the Earth may become five degrees Fahrenheit warmer by the year 2050. This could be enough to melt the polar ice caps and glaciers, at least during summer, and flood coastal cities like New York, Boston, Miami, and Los Angeles. Ten percent of our world's total surface area is covered with ice. If this ice were to melt completely, sea levels would flood more than half the land mass of planet Earth. In August 2000 a patch of open water about one mile wide appeared at the North Pole, where the ice is usually six to nine feet thick. Although records show that this has happened before and is not necessarily in itself a danger sign, it is one of many indications of an ongoing warming trend. In February 2001, for example, Dr. Lonnie G. Thompson, a senior research scientist at Ohio State University's Byrd Polar Research Center, reported that the snows of Mount Kilimanjaro, which have been a constant feature of the central African landscape for thousands of years, are melting away and could be entirely gone within fifteen years.

Greenhouse gases are the primary cause of global warming. As they accumulate in the atmosphere, they reflect heat back to Earth and turn the entire planet into a very undesirable sort of greenhouse. The main culprit gases are carbon dioxide from burning fossil fuels and wood; methane from biogas, bacterial decomposition in the guts of livestock and other animals, and rice farming; and chlorofluorocarbons (CFCs) from freon, a refrigerant. Carbon dioxide probably accounts for fifty percent of the rise in global temperature, and the United States accounts for twenty-five percent of the world's total carbon dioxide emissions while possessing only four percent of the world's population. Burning fossil fuels alone releases about 2.5 billion tons of carbon dioxide a year, of which two thirds are emitted into the atmosphere (the remaining third is absorbed by the oceans).

Reducing the buildup of these gases is our best **Protective Response Strategy** against global warming. CFCs have now been banned in most developed countries, but they are still being produced in large quantities in developing countries in the Third World. Carbon dioxide emissions could be cut significantly by switching to natural gas, solar and geothermal energy, and nuclear energy that is truly clean instead of Chernobyl dirty. Banning clear-cut timbering and working to replant forests would also help, because forests play an important supporting role to microscopic algae floating on the oceans in transforming carbon dioxide into oxygen via photosynthesis. In June 2001, President George W. Bush suggested that carbon dioxide capture, storage, and sequestration technologies are viable alternatives to conservation efforts. While promising, the commercial viability of these technologies is at least a decade away. We should also try to develop supplements for livestock feed that will reduce the flatulence of cattle which, believe it or not, releases thirty million tons of methane into the atmosphere annually. Adopting practices that allow for periodic draining of rice paddies will also reduce methane in the atmosphere. Methane burns cleanly as fuel, however, so as I mentioned earlier in the book it would also be desirable to employ methane-producing microbes in a controlled way to satisfy some of our energy needs. There is also a great untapped supply of methane below the ocean floor.

Clean nuclear energy and methane from microbes are achievable prospects, but they must await future technological innovations. One step we could take now is to roof buildings with more reflective, light colored materials, especially in cities. The unrestrained growth of megacities around the world, such as greater

Los Angeles, Mexico City, and São Paolo, to name a few, adds considerably to global warming. Urban structures can become 50 to 70 degrees Fahrenheit warmer than the surrounding air, and this can make the ambient temperature of cities ten degrees warmer than that of the countryside.

Global warming obviously has many more ramifications than its impact on the mosquito population. But the connection between the two illustrates the complexity of our interactions with the environment and the pressing need for us to turn to ways of working with nature rather than against it. Mosquitoes and malaria have proven themselves to be far too resilient and adaptable to be defeated with crude tactics, and it is high time that we tried to devise subtler and more effective approaches. If we continue heedlessly to transform the environment to meet our short-term goals, we will never know when we are going to become vulnerable to another virus like WNV or HIV by releasing it from its traditional environment and spreading it around the world in airplanes and other ways. Warmer temperatures will eventually broaden the reach of disease-carrying arthropods and rodents, thereby increasing the incidence of infectious disease.

In this regard it's worth considering whether new germ threats may learn to coopt mosquitoes in order to spread among human beings. For the present it seems that HIV cannot be transmitted by mosquito bites, because the virus is not able to invade the insect's salivary glands and is completely broken down in its digestive tract. But with a virus that mutates as rapidly as HIV, who knows what the future may hold? It used to be thought that hepatitis C virus (HCV) could be transmitted only via blood-to-blood or sexual contact, just like HIV. The Pasteur Institute in Paris recently reported that this may not be so. Scientists there have demonstrated that HCV can bind to and replicate in mosquito cells, which would be a prerequisite for transmission. These researchers suggest that mosquito transmission could account for up to twenty percent of HCV infection. This is profoundly significant, because hepatitis C is the most common blood-borne infection in the United States. It is currently estimated that as many as four million Americans have antibodies to HCV, indicating ongoing or previous infection. These individuals are at elevated risk for chronic liver disease, and liver disease caused by HCV is the most common indication for liver transplants among adults.

On Eight Little Legs

Two sisters, ten and seven years old, spend most of a long summer weekend playing with their dogs in the woods and grassy fields near their home in upstate New York. The children know that the area abounds with ticks, and their parents have taught them to check themselves carefully for ticks, or the sign of a tick bite, at the end of the day, because ticks can be a vector for many infectious germs. On Sunday evening the ten-year-old notices a large bull's-eye rash on her trunk. A fever develops overnight, and the girl's parents take her to the pediatrician, who diagnoses Lyme disease transmitted by a tick bite and prescribes antibiotics.

Episodes like this have become common since Lyme disease was first discovered in Connecticut in 1976. Subsequently it was recognized that Lyme disease occurs throughout much of the United States as well as in several other countries. If mosquitoes are the number-one insect vector for infectious diseases, ticks easily take second place. Ticks are bloodsucking parasites in their own right, just like mosquitoes, but they move about on eight little legs instead of wings. They may be black, brown, or brightly colored, and some of them are microscopically small. Depending on their species and habitat, ticks can transmit viruses such as WNV and SLE or parasitic bacteria that cause such illnesses as Q fever, relapsing fever, Rocky Mountain spotted fever, and tularemia ("rabbit fever"), in addition to Lyme disease. These varied bacteria-induced fevers and Lyme disease share many of the same symptoms: headache, fever, chills, cough, muscle aches, fatigue, and in severe cases central-nervous-system effects that may progress to delirium and coma. All may be fatal, and all require speedy treatment with antibiotics.

The overlap of symptoms made it hard to recognize Lyme disease as a distinct illness. In fact, the medical profession sometimes refers to Lyme disease as "the great imitator," because it mimics so many other illnesses. Yet in the summer of 1976 a cluster of cases in Lyme, Connecticut, caught the attention of Dr. Allen C. Steere, a Yale University rheumatologist, who described the disease and named it for the town. Six years later, in 1982, Drs. Willie Burgdorfer and A. G. Barbour confirmed that, despite all its similarities to other ailments, Lyme disease is a distinct illness with its own characteristic disease process. Burgdorfer and Barbour isolated the specific cause of the illness, a previously unrecognized spirochete (that is, spiral-shaped) bacterium now called *Borrelia burgdorferi,* from the body of the tick *Ixodes ricinus.*

Ixodes ticks are unusually tiny. The larval stage is the size of a grain of sand, the nymph is the size of a poppy seed, and the adult is only the size of a sesame seed. Even the adult's bite is so small that it often goes unnoticed. When *Ixodes* ticks take a blood meal from an animal, they can regurgitate a dose of *Borrelia burgdorferi.* In sixty to eighty percent of human victims, the first sign of infection is a rash, usually but not always at the site of the tick bite and usually but not always resembling a bull's-eye target. Because the rash can migrate to any part of the body, it is designated erythema migrans (EM). It can last for days to weeks, followed by joint pain, which may be the harbinger of chronic arthritis, nervous-system disorders, and/or heart trouble. All of these symptoms may come and go, greatly complicating diagnosis.

Lyme disease must have existed for an unknown length of time before scientists analyzed and named it. Why was it discovered only in 1976? The short answer is that tick-borne illnesses, like mosquito-borne illnesses, wax and wane in response to the direct or indirect impact of human actions. By the mid-1970s, human beings' collective behavior had created circumstances that so favored the spread of Lyme disease that sooner or later it was bound to attract people's notice and demand a protective response. As so often happens with human beings and infectious illness, we made our own sickbed and then we had to lie in it.

Although the *Ixodes* ticks that transmit *Borrelia burgdorferi,* the cause of Lyme disease, are also known as deer ticks, this label is a little misleading insofar as the illness is concerned. *Ixodes* larvae feed on white-footed mice, and these mice are the main hosts and reservoirs of *Borrelia burgdorferi.* The tick nymphs and adults feed on deer and incidentally on people. A key predictor for the incidence of Lyme disease from one year to another is thus the acorn harvest, which supports both mice and deer. More acorns mean more white-footed mice and more deer, and consequently more ticks and more Lyme disease. If we look at the whole chain of infection involving bacteria, ticks, mice, deer, and people, we can see that the single greatest factor in the extent of the disease is the extent of forested areas near where people live. And the extent of forested areas depends on human choices and actions.

When the Pilgrims formed the Mayflower Compact on the shores of Cape Cod, Massachusetts, in 1620, most of the continent from the Atlantic Ocean to the Mississippi River was covered by dense forests that were home to millions of deer and other wildlife. The Pilgrims and the European settlers who followed them saw the forests as the enemy, to be cleared root and branch for farmland. They were so successful that by 1854, when Henry David Thoreau published *Walden,* his meditation on people and nature, most of the same territory was open farmland and pastureland. And the deer were nearly extinct. They could survive in dense forests in a fluctuating balance with their pre-Colonial predators: wolves, cougars, coyotes, and Indian hunters without firearms. When the forests were timbered and burnt out, however, they had no place to hide from white and Indian hunters armed with rifles.

In the century and a half after *Walden's* publication, land-use patterns in the eastern United States changed dramatically. The economy of New England and the Middle Atlantic states became predominantly industrial and commercial, rather than agricultural. One wonderful benefit of this change is that the forests came back, a development that started slowly and that has accelerated dramatically in the last few decades. There is now more forest in New England, for example, than at any time since shortly after the first European settlers arrived.

As the forests regrew, the populations of deer and other animals also rebounded, and there are now once again millions of deer in the eastern United States, a substantial portion of the approximately twenty-five million deer in the country as a whole. Today's deer face little or no threat from animal predators, however, and human hunters can only cull a small fraction of them each year. The result is that deer now roam very much at will throughout eastern forests and the often heavily wooded suburbs of New England and the Middle

Atlantic states. As many a gardener can unhappily testify, deer love to eat the tender shoots and leaves of many show plants and vegetables. In most suburban communities in the eastern United States, deer rank at the top of the list of garden pests, harder to deter and defeat than virtually any insect one can name. And, of course, as the deer forage, they bring Lyme-disease-transmitting ticks into the neighborhood with them to infect mice and people alike.

Ironically, before the deer–Lyme disease connection was made, one community tried to turn the resurgent deer to its advantage in what I call the Block Island Blunder. Block Island, twelve miles off the coast of southern Rhode Island and within ferry distance of Connecticut, Massachusetts, and eastern Long Island, was once a fishing and farming community. There were no deer on the island, because it was too far from the mainland for deer to swim there. As the economic viability of local fishing and farming faded in the twentieth century, the residents of Block Island fell on hard times. Summer tourism substituted for some, but not all, of the loss. So community leaders decided to import deer from the mainland with an eye toward serving hunters in the fall months. But with the deer came *Ixodes* ticks and *Borrelia burgdorferi* bacteria. Today Block Island has become a hot spot for Lyme disease; the deer are less a draw for hunters than a public-health nuisance.

As if Lyme disease were not enough on its own, *Ixodes* can also transmit two other potentially serious illnesses, Ehrlichiosis and babesiosis. Some victims of *Ixodes* tick bites actually come down with two or three ailments at once. Ehrlichiosis is caused by the bacterium *Ehrlichia,* which is related to the varied tick-borne bacteria known as *Rickettsia* that cause such diseases as typhus and Rocky Mountain spotted fever. The *Ehrlichia* bacteria infect the white blood cells of human beings, sometimes initially mimicking a mild hepatitis with fever and elevated liver enzymes. The symptoms of Ehrlichiosis include fevers as high as 105 degrees Fahrenheit, unrelenting headaches, chills, and vomiting. On rare occasions, the disease can lead to the failure of internal organs and death. An outbreak of Ehrlichiosis recently occurred among residents of a golf-oriented retirement community that abuts a natural wildlife preserve in Florida. While the retirees were out on the golf course, they were being preyed upon by *Ixodes* ticks.

Babesiosis, for its part, is caused by a protozoan parasite called *Babesia microti* that infects red blood cells in much the same way the malaria parasite *Plasmodium* does. Its symptoms include low-grade fever, sweating, fatigue, and malaise, and it, too, can be fatal. Fortunately babesiosis is very rare in human beings. Only a few dozen cases are reported annually nationwide, as opposed to about two thousand for Ehrlichiosis and about thirteen thousand for Lyme disease.

In 2001, researchers announced that *Ixodes* and another tick species, *Amblyomma* (or the Lone Star tick, because it was first identified in Texas), can transmit *Bartonella,* the germ that causes cat scratch fever. This explains cases of the disease that have no relation to its usual sources, dogs and cats or their fleas. And it partly fulfills an educated guess I made in 1998, when I wrote a chapter for a medical text on *Bartonella* and suggested that a number of arthropods, like insects, ticks, and spiders, were probably involved in its transmission. But the real significance of the finding is that it shows, once again, that we still have an enormous amount to learn about the place of germs in the planetary ecosystem and in human health. No doubt there are many microorganisms like *Bartonella* that are ubiquitous in nature and that play as yet unrecognized roles in causing disease.

The impact of infectious germs whose vector is ticks will continue to wax and wane in response to the behavior of human populations. But unlike germs whose vector is mosquitoes, they are unlikely to become a big enough threat to require large-scale public-health responses. That leaves a **Protective Response Strategy** mainly up to the individual. An effective **Protective Response Strategy** for tick-borne illnesses has two prongs, prevention and treatment.

PREVENTION:

- **Wear long pants and long-sleeved shirts when you spend time in or near wooded areas during tick season, which lasts from April to October. Light-colored clothes are best, because ticks will show up against them more sharply.**

- Apply DEET, a tick repellent, to exposed skin, and/or spray clothing with premethrin products, which kill ticks.

- At the end of a stay outdoors, examine yourself and children carefully for evidence of ticks. Check especially between fingers and toes; behind the knees; in groin, armpits, bellybutton, and ears; and around the neck, hairline, and the top of the head. Ticks prefer the warmth of any little crevice or bend in the body.

- If you find a tick, don't panic. It takes a minimum of twenty-four hours for a tick to settle in an area of the body and transmit disease. The time lag is mostly owing to the fact that the tick seeks out a favorable spot on the body before it bites. In addition, any germs inside the tick remain dormant until the tick has taken a blood meal.

- In the case of Lyme disease, the *Borrelia burgdorferi* spirochetes respond to the victim's blood by changing their outer surface proteins (OSP), so that they can then invade the gut wall of the tick and spread throughout its tissues. This process culminates in the tick's secreting *Borrelia* germs into the cells of the animal it is biting.

- Ticks should never be removed with bare hands. Use a tweezers or a tissue to remove the tick by the head first. If the tick is biting into the skin, don't try to pull it off. Its head will stay attached and can still transmit infectious germs. Sometimes you can make a tick release its hold by bringing a hot needle or a match near its rear, almost touching it. Blow the match out first, of course, to avoid burns.

- If you find and remove a tick before it seems to have bitten you, stay on the alert for any signs of rash or other symptoms for two to four weeks. As far as Lyme disease is concerned, keep in mind that ten to fifty percent of ticks in the eastern United States carry *Borrelia burgdorferi*, but only two percent of ticks in the western part of the country do so.

- If you live in a region where Lyme disease is very prevalent and you enjoy spending time gardening or in other outdoor activities, consider being vaccinated with Lymerix, which provides protection against Lyme disease in about eighty percent of the people who take it. Lyme disease has been found in forty-five states and the District of Columbia, but the overwhelming majority of cases so far come from only ten: Connecticut, Delaware, Maryland, Massachusetts, Minnesota, New Jersey, New York, Pennsylvania, Rhode Island, and Wisconsin.

  The Lymerix vaccine works by cleverly taking advantage of the time lag between a tick's biting its victim and the transmission of *Borrelia* bacteria. When the *Borrelia* move from a tick into a mammal's body, whether mouse, deer, or person, they change their OSP in an effort to fool the new host's immune system. Inside the tick *Borrelia* wear, so to speak, OSP-A; inside human beings they wear OSP-C. This shifting is a biologic trick that scientists call "down regulation" of one protein in favor of another. The Lymerix vaccine is designed to stimulate antibodies against OSP-A, so that when the tick takes a blood meal the antibodies in that blood can neutralize the *Borrelia* bacteria inside the tick, preventing them from ever transferring into a person's body. New evidence suggests that additional outer surface proteins are involved in infection, and this knowledge should allow the development of a more effective vaccine.

- If you want to try to keep deer away from your house and garden, you're taking on a frustrating challenge. Some people report success with soap and with male human urine; deer don't like the smell of either substance. But

hungry deer will usually disregard their aversion and keep on coming.

TREATMENT:

- **If you find that a tick has bitten you, or if you experience any symptoms of tick-borne illness, consult a physician as soon as possible. It is imperative that a speedy diagnosis be made so that the correct antibiotic can be administered to prevent long-term health problems. Tick-borne illnesses that are not treated promptly may result in chronic relapses and multiple organ damage or failure.**

  Lyme disease in particular requires quick, accurate diagnosis, because the _Borrelia_ bacteria can shed their cell walls and assume what is known as an L-form. The term L-form comes from the Lister Institute in Great Britain, where the phenomenon was first observed. Many antibiotics, notably penicillin, work by attacking a germ's cell wall, and these will be useless against _Borrelia burgdorferi._ Assuming L-form allows _Borrelia_ to slip and slide in between body tissues and remain safe from antibiotics, unless the right ones are prescribed and administered.

Borrelia's adaptability is yet another example of the astonishing variety of strategies that germs have employed in their approximately four-billion-year history on Earth. Those strategies have played a significant, often tragic role in our own history on the planet, as we've seen. Now it's time to turn to what the future holds as germs and people continue to coexist on Earth, and as germs promise to play an ever greater part in medicine on the one hand, and in warfare and terrorism on the other.

PART III

The New Age

The mystery of life does not reside in its
manifestation in adult beings, but rather
and solely in the existence of the germ
and its becoming.

—Louis Pasteur, 1860

10

Not the Usual Suspects

The more science discovers about germs, the more roles they are found to play both in keeping people healthy and in causing diseases. At the most fundamental level, germs, or germ vestiges, seem to be essential to the healthy functioning of the cells of all higher life-forms. And evidence increasingly points to infectious germs as triggers and contributory causes for illnesses, such as cancer and heart disease, that no one ever suspected them of being involved in before. Consider the case of ulcers.

Man on a Mission

In 1983, Dr. Barry Marshall, a young research physician in Perth, Australia, decided to make himself sick. It was the most dramatic and conclusive way he could think of to prove that the worldwide medical establishment was wrong, and that he was right, about the cause and proper treatment of peptic ulcers.

Although popular entertainment has often presented the milk-drinking ulcer sufferer as a figure of fun, a peptic ulcer is no laughing matter. An ulcer is an erosion of skin or other body tissue. Peptic, or gastrointestinal, ulcers occur when gastric juices—hydrochloric acid and digestive enzymes, especially one called pepsin—eat away at the lining of the stomach or the duodenum, the upper part of the small intestine. The ulcerations appear as round or oval lesions in the duodenum near the juncture with the stomach wall and in the stomach wall itself. Such lesions can obviously play havoc with a person's digestion, leading to vomiting and other forms of gastrointestinal distress. But the main symptom is burning pain one to three hours after eating. In extreme cases peptic ulcers can burn a hole right through the stomach or intestinal wall, producing internal bleeding and even death.

Ulcers are a common malady. They afflict three to eight percent of the population in the Third World, and .5 to one percent of the population in developed countries. They were long thought to be caused by heredity, diet, traumatic stress such as burns or complications after surgery, and even chronic anxiety. To treat them, doctors advised changes in diet to less acidic and more alkaline foods like milk, and prescribed medicines like Zantac and Tagamet that ease symptoms by modulating acid production in the stomach. In extreme cases surgeons could operate to repair holes or remove damaged portions of the stomach. Surgery might also involve severing the lower end of the vagus, or "wandering," nerve. The longest of ten cranial nerves, the vagus nerve runs from the brain to the abdomen and carries signals to and from the ears, pharynx, larynx, and vocal cords as well as the stomach. Cutting off its lower end decreases stimulation of gastric juices and so can provide some relief of symptoms in the worst ulcer cases. It's not a cut to make lightly, however, because postoperative complications can include severe diarrhea, weight loss, and malnutrition. In fact, all these treatments have a fatal flaw: They can help manage and ease the chronic suffering from ulcers, while creating some problems of their own, but they cannot cure the underlying disease process. Although the body does heal some ulcers, the prevailing medical dogma in 1983 was that most people with ulcers simply had to learn to live with them.

Barry Marshall was convinced that all this could change. His research mentor, Dr. Robin Wright, a pathologist at Royal Perth Hospital, had for some time been finding an unclassified bacterium in stomach biopsies. This was puzzling, because the stomach was thought to be free of germs, except for those that are harbored transiently with food and quickly passed into the intestine. Marshall and Wright identified the previously unrecognized bacterium as *Campylobacter pylori*—in 1989 it was reidentified as belonging to a separate but related genus and named *Helicobacter pylori*—and Marshall inferred that it was the cause of peptic ulcers and that treatment with antibiotics could kill the bacterium and cure ulcers.

This set the stage for Marshall to experiment on himself in accordance with Koch's Postulates, the classic framework for identifying a particular germ as the cause of a particular disease. Marshall first examined a number of people with ulcers and found that they all harbored *Helicobacter pylori.* He then had himself checked for both ulcers and *H. pylori.* Once it was established that he had neither, he drank a cocktail of *H. pylori.* About a week later he began to feel sick. An examination showed that he was developing an ulcer and *H. pylori* was isolated from his system, proving the link between the disease and the bacterium. Finally, treatment with antibiotics knocked out both the ulcer and the *H. pylori.*

Marshall's brilliant stunt did not change medical practice overnight, as he'd hoped it would. It was not just that doctors were used to treating ulcers in a different way and resisted giving up old habits, or that the pharmaceutical companies wanted to protect a multibillion-dollar market in ulcer medications. To be sure, both these factors posed significant barriers to the adoption of Marshall's theory. But at a more basic level, medical science needed to figure out how a bacterium like *H. pylori* could survive in the stomach long enough to set a disease process in motion.

As the story of carrion avoidance in Chapter 7 illustrated, the stomach is a harsh environment for infectious germs. In human beings the stomach is washed with half a gallon of gastric acid every day, and the interior of the stomach has a highly acid pH of 1 to 2. This is more than acidic enough to kill *H. pylori,* which prefers a neutral pH of about 7. In addition, the digestive process in human beings quickly transfers ingested food and other matter from the stomach to the intestines, where the body's own enzymes and the normal gut flora break down food, extract nutrients, and form feces.

These factors made it seem as if Marshall's theory could not be right. But by the mid-1990s, the work of many different researchers, including Marshall, led to a new understanding of the microbiology of the stomach. It turns out that *H. pylori* can burrow under the mucous coating of the stomach lining like a secret agent going deep under cover as a mole in the enemy's camp. Underneath the mucous coating, the pH is a welcoming 7.4, perfect for *H. pylori*. The immune system detects the presence of intruding bacteria and repeatedly sends T cells to wipe them out, but the T cells are stymied by the mucous coating and they accumulate futilely on one side of it, along with other immune-system cells. Eventually they die, spilling their cell-killing contents onto stomach lining cells instead of *H. pylori* and initiating an inflammatory process. Meanwhile the *H. pylori* bacteria feed on the nutrients the body sends to support the T cells. They also release urease, which converts the saliva and gastric acid in the stomach to bicarbonate and ammonia, neutralizing the pH and enhancing their own growth. The inflammatory immune response and the growth of *H. pylori* reinforce each other, and this can result in chronic active gastritis, which in turn can lead to the formation of ulcers. *H. pylori* has now been linked to ninety percent of duodenal ulcers and sixty to seventy percent of gastric, or stomach, ulcers. Most other ulcers stem from the use of aspirin and other anti-inflammatory drugs to treat arthritis and other ailments.

In Third World countries about eighty percent of the population harbor *H. pylori* by age twenty. In developed countries like the United States, the prevalence of *H. pylori* infection increases one percent for each year of life, meaning that about thirty percent of thirty-year-olds have it and about sixty percent of sixty-year-olds, or about forty percent of the adult population overall. Among Native Americans in Alaska, and no doubt in some other poor communities in the United States, the incidence of *H. pylori* infection rises to Third World levels. Because *H. pylori* is an invasive intestinal germ that is passed in feces, the main route of transmission from person to person is thought to be by the fecal-to-oral route. But *H. pylori* can also be cultured from dental plaque, and so the oral-to-oral route also seems to be responsible for some transmission. The potential for *H. pylori* to be present in dental plaque and stomach contents means that it can sometimes be detected by a breath test. Recently researchers have reported that *H. pylori* frequently occurs in well water. The greater reliance in poor rural areas on well water rather than properly filtered municipal drinking water may account for some of the greater incidence of *H. pylori* infection in the Third World and in some American communities. Interestingly, *H. pylori* antigens were detected in 1,700-year-old pre-Columbian mummies, and current evidence indicates that they are nearly universally found in primate stomachs (stomachs that are at least 400 million years old).

An incidence of eighty percent among Third World adults and forty percent among adults elsewhere makes *H. pylori* one of humanity's most common bacterial infections. As with other germs, the rates of overt infection and full-blown disease are, of course, much lower. Again, three to eight percent of Third World adults develop ulcers as a result of *H. pylori* infection, and .5 to one percent of adults in other countries do so. This indicates that ulcer formation is multifactorial, scientific jargon for having multiple causes. In addition to *H. pylori,* psychological conditions, diet, and genetic abnormalities or predispositions may all be involved. For example, a diet rich in very acidic foods may facilitate the formation of ulcers, and so may a somewhat slow digestive process, which keeps acidic foods in the stomach longer than usual.

Thanks to Barry Marshall and other researchers, we now recognize that the multifaceted problem of ulcers requires multi-pronged treatment. Antibiotics must lead the way, although formulating an effective antibiotic cocktail for *H. pylori* can be tricky, because many antibiotic-resistant strains have emerged. Changes in diet and lifestyle may help to keep reinfections with *H. pylori* from causing problems. And there is still a place in ulcer treatment for antacid compounds like Zantac and Tagamet, even as their manufacturers have cannily repositioned them as nonprescription remedies for indigestion and heartburn. For severe ulcers, surgery continues to be a last resort. But the bottom line is that most ulcers can now be cured, rather than just treated to reduce symptoms.

Although it was not recognized at the time, there was supporting evidence for Marshall's theory about *H. pylori* and ulcers in the late 1970s and early 1980s. For example, Dr. Martin J. Blaser, then conducting basic research at Vanderbilt University School of Medicine, suggested that bacterial cytotoxins might be involved in peptic ulcers, and he noted that such toxins are produced by about half the strains of *Campylobacter* bacteria, the group that was thought to include *Helicobacter* until new genetic techniques led to its being reclassified in a different, but closely related, genus. Over the last two decades, scientists have discovered that these common intestinal invaders have many surprises in store for us. One of the most important to date is that *H. pylori*–induced gastritis not only leads to ulcers, but also helps initiate stomach cancers, a link that Martin Blaser, now chairman of medicine at NYU School of Medicine, is helping to map. The World Health Organization (WHO) has labeled *H. pylori* a Type 1, or "causal," carcinogen.

Hidden Plagues

Another breakthrough is the finding that chronic infection with *H. pylori,* indeed chronic infection with almost any germ in almost any part of the body, from gum disease to urinary-tract infections, can set the stage for heart disease. What happens is that the germs make their way into the bloodstream, where they can stimulate clotting factors that block coronary arteries. Or the germs themselves can accumulate in such high concentrations that they clog the arteries like the scum on the inside of a pipe. For example, there seems to be a clear interaction between high cholesterol and the inflammation caused by germs. If germ growth causes more immune cells to be present in arteries, the immune cells can latch on to cholesterol particles, resulting in plaque and clogging of the arteries.

The idea that bacteria can trigger cancer and heart disease was until very recently as laughable to most scientists and physicians as Barry Marshall's insistence that they can cause ulcers. One germ, one disease, one treatment has long been the mantra of medical textbooks and protocols. But almost every day new evidence shows that one-dimensional orthodoxies are obsolete, and that infectious germs often have a multidimensional impact that calls for a multidisciplinary approach in both research and treatment. The problem is usually not so much that a prevailing orthodoxy is flat-out wrong, but that it is incomplete, just as the standard explanation of ulcers and stomach cancers was incomplete until the role of *H. pylori* was recognized.

Many diseases that have long plagued humanity now seem to be caused, at least in part, by germs. For example, researchers have established links between infectious germs and several types of cancer, adding significantly to our understanding of how many different factors, including germs, combine to trigger cancerous growths.

Hepatitis B and C viruses have been linked to the development of hepatocellular carcinoma, liver cancer, that may appear fifteen to sixty years after the onset of chronic infection.

Herpes viruses, including herpes simplex 2, have also been associated with cancer. Human herpes virus-8 (HHV-8) has been extracted from Kaposi's sarcoma tumors. Kaposi's sarcoma is a form of cancer that is common in late-stage AIDS and that affects the skin, the lymph nodes, and the internal organs, particularly the gastrointestinal tract and the lungs. The presence of antibodies for HHV-8 can now serve as an early-warning sign of potential Kaposi's sarcoma. The latest research shows that deep kissing may spread HHV-8; in most people the virus does not help trigger cancer until the immune system is affected by other factors, such as infection with HIV.

Still another herpes virus, Epstein-Barr, which causes mononucleosis, has been found to cause cancer of the nasopharynx, the upper part of the throat, and the lymph system in some population groups. Epstein-Barr-related nasopharyngeal cancers are particularly common in China's male population. Likewise, one of the commonest cancers in Africa is Burkitt's lymphoma, which results in tumors of the jaw in children and young adults. Epstein-Barr virus has been found in ninety percent of these tumors. The malaria parasite, *Plasmodium,* seems to be a co-factor in these lymphomas, an intriguing example of infectious germs interacting with one another. Heredity and environment surely play contributing roles, as well.

Human papillomavirus (HPV) has long been recognized as the source of genital warts, one of the world's most common sexually transmitted diseases (STDs). In the United States over five million new cases of HPV infection are estimated to occur annually. The warts, which can look like tiny cauliflower buds, appear on the anus in both men and women, on the labia and vulva in women, and on the scrotum and the tip of the penis in men. If a person performs oral sex on someone infected with HPV, lesions in the mouth and throat can result.

As if these effects weren't serious enough, HPV infection has emerged as the leading cause of cervical cancer. Investigators have speculated that it may also be responsible for fifteen percent of oral cancers, especially cancers of the tonsils. There are fifteen thousand new cases of cervical cancer in the United States every year, and five thousand deaths. Worldwide, cervical cancer kills about 250,000 women annually, with women from developing countries accounting for eighty percent of the total. To check for cervical infections that might lead to cancer, ob-gyns apply vinegar, which causes infected areas to whiten. If identified early enough, most cervical cancer can be effectively treated and cured.

This will save the lives of men as well as women, because men can apparently "catch" cancer from women with cervical cancer. Swedish researchers recently studied seven thousand men whose wives had cervical cancer. They learned that the men whose wives had very invasive forms of the disease, about half the group, had twice the normal rate of anal cancer and one and a half times the normal rate of penile cancer, as well as increased risk of leukemia and esophageal, pancreatic, and lung cancer. Related research indicates that uncircumcised men have a slight but statistically significant added risk of contracting, and thus passing on, HPV infection. Muslim men, for example, are circumcised in their youth, and there are somewhat fewer cases of cervical cancer among Muslim women. Uncircumcised men thus need to take greater care with this aspect of personal hygiene; but whether a man is circumcised or uncircumcised, a condom is still called for in any sexual activity outside the boundaries of a monogamous relationship.

The traditional test for cervical cancer has been a Pap smear, named for the Greek-American scientist George Papanicolaou, but the results are frequently ambiguous. An HPV test can now clear up matters. If a woman has an abnormal Pap smear and then tests positive for HPV, she faces a 116-fold risk that she will develop high-grade cervical lesions, perhaps as a prelude to cancer. If she has an abnormal Pap smear, but tests negative for HPV, the strong likelihood is that she will soon return to normal.

More recently, the bacterium *Chlamydia trachomatis* has also been linked to anorectal and cervical cancer. This connection may have been a little slower to emerge because infection with *C. tra-*

chomatis often goes unnoticed, even though it is the number-one bacterial STD in the United States. Three million men and women in the United States are infected with *C. trachomatis* every year, and rates of infection are soaring throughout the Western world, especially in women under the age of twenty-five. But in many cases no symptoms develop, and in others *C. trachomatis* infection coincides with, and can be masked by, gonorrhea. In women, *C. trachomatis* can lead to pelvic inflammation, ectopic pregnancy, and infertility. Only about fifteen percent of infected women experience symptoms, which can include vaginal bleeding and abdominal pain. In men, the germ can cause discharges from the penis and painful urination, among other problems, and about sixty percent of infected men develop symptoms.

Researchers at the University of Helsinki and Finland's National Public Health Institute conducted a study of 530,000 Finnish, Norwegian, and Swedish women. They learned that two strains of *Chlamydia trachomatis* increase the risk of contracting cervical cancer three times and four times, respectively, and that women infected with the "G" strain of *C. trachomatis* are about seven times more likely than normal to contract cervical cancer. Combinations of these strains present a still greater risk.

A closely related germ, *Chlamydia pneumoniae,* has been identified as a contributor to heart disease. *C. pneumoniae* bacteria cause pneumonia, bronchitis, sinusitis, sore throats, and earaches. But they can also invade arterial cells and make them foam out their insides to form atherosclerotic plaque. Some scientists have suggested that *C. pneumoniae* infections of coronary blood vessels will most often occur when iron levels are high. This is an attractive idea because it would help explain why premenopausal women, who lose iron in menstrual blood, have a lower incidence of heart disease than men do. After menopause, as stored iron levels rise, the incidence of heart disease among women rises to match that of men. Exercise can lower stored iron levels, and thus may lower the risks of *C. pneumoniae* infection for men and women alike.

The occasion for more than ten million physician visits a year in the United States alone, urinary-tract infections (UTIs) can be caused by a number of infectious germs, including *E. coli,* a part of human beings' normal flora. Another common intestinal bacterium, *Proteus mirabilis,* can also cause UTIs, but usually only in people with an anatomical fault or those who must use an indwelling urinary catheter. The kicker is that *P. mirabilis* can cause bladder or kidney stones to form from magnesium ammonium phosphate ("struvite") or calcium phosphate ("apatite"). Normal urine is highly acidic. But like *H. pylori, P. mirabilis* can release the enzyme urease, which in this case splits urea into ammonia and carbon dioxide, raising the pH and making the urine more alkaline. The calcium and magnesium salts in urine can then precipitate out and form stones. Discovery of this process could well be the first step to a preventive vaccine that will prevent *P. mirabilis* from attaching itself to bladder cells.

Skepticism initially greeted all these surprising germ-disease connections. But as more and more of them have been identified, they have spurred medical detective work to advance further along the same lines. Many scientists have long suspected, for example, that a viral infection of some kind might play a role in amyotrophic lateral sclerosis (ALS), the degenerative central-nervous-system disorder that is also known as Lou Gehrig's disease, because that great New York Yankee of the 1920s and 1930s remains its most famous victim. A recent study has identified a candidate virus, echovirus-7, which has previously been shown to cause meningitis and encephalitis. Researchers found echovirus-7 in the damaged nerve cells of fifteen of seventeen ALS patients. By contrast they found the virus in only one of twenty-nine control patients. Although the number of people studied so far is small, the almost ninety percent correlation of echovirus and ALS represents perhaps the most promising avenue yet for developing a cure for this incurable and always fatal disease.

Finding new germ-disease connections depends on increasingly powerful genetic and molecular techniques that allow scientists to look inside human cells and observe them in a multitude of interactions with infectious germs and other substances. These techniques may soon make it possible to settle such long-standing controversies as whether silicone breast implants, which often leak into surrounding tissues, can cause a lupus-like autoimmune disease. Most medical authorities have dismissed such claims as groundless. But recent experiments have proven that silicone gels and oils can indeed stimulate the production of autoantibodies and other autoimmune phenomena that are typical of lupus. In the spring of 2001, researchers at the National Cancer Institute reported that they had found higher than normal rates of lung and brain cancers in women with silicone breast implants.

A speculative finding that needs to be confirmed or refuted by more research is that some cases of schizophrenia, depression, and obsessive compulsive disorder may be connected to Borna disease virus (BDV). Borna disease ordinarily attacks the central nervous systems of horses and other vertebrate animals, and manifests itself in behavioral abnormalities. BDV is an unclassified RNA virus, similar to the rabies virus, but it is unique in the way it replicates in the nuclei of the cells it invades. Antibodies to BDV have been found in some people with neuropsychiatric disorders such as depression, schizophrenia, and obsessive compulsive disorder. But no one can yet say for certain whether BDV can actually cause or trigger these disorders in some way. Japanese scientists have isolated BDV from the brain of a deceased schizophrenic patient, and they have shown that the virus can successfully establish itself, and cause disease, in the brains of gerbils. A more recent study at Fukushima Medical University in Japan found no significant link between BDV and neuropsychiatric disorders. But a team of researchers at the Johns Hopkins University School of Medicine reported in the spring of 2001 that a group of patients with schizophrenia had high levels of a retrovirus called HERV-W. More studies will obviously be needed to determine what role, if any, viral agents and other infectious germs play in devastating mental illnesses like schizophrenia.

Equally fascinating is the possibility that Alzheimer's disease may be caused by viruses such as herpes simplex 1, the virus which gives people cold sores, or by subviral particles like the prions involved in Creutzfeldt-Jakob disease (CJD). In laboratory experiments, herpes simplex 1 killed brain neurons and covered them with the distinctive beta-amyloid plaque that has been found in tangles of dead neurons in the brains of Alzheimer's patients. Similar plaques and tangles have been found in CJD patients, suggesting a possible common mechanism of viral infection. Attempts to verify this by using brain samples from people who have died from Alzheimer's disease in order to transmit the disease to primates and rodents have not so far been successful.

Not every speculative theory about germs will pan out, to be sure, but two more examples will, I hope, illustrate the value of pursuing this sort of research wherever it leads.

You Are What You Eat

Inflammatory bowel disorders, syndromes, and diseases (IBDs) afflict tens of millions of people to varying degrees. Their symptoms run the gamut from unpleasant, perhaps embarrassing bouts of gassiness and flatulence to ulcerative colitis and Crohn's disease. A chronic severe inflammation of the ileum, a portion of the small intestine, Crohn's disease, also known as ileitis, may damage the liver, skin, or eyes, bring on arthritis, and even kill.

When Dr. B. B. Crohn identified the disease, in 1932, he suggested that infectious microorganisms might be responsible for it. That has remained an open possibility ever since. Many candidate germs have been put forward, especially in recent years: *Peptostreptococcus, Listeria monocytogenes, Campylobacter jejuni, Salmonella, Shigella, Yersinia,* and *E. coli,* among others. Some are members of the normal gut flora, if not always the usual strains of them, and others are invasive germs seeking to colonize new territory. But no one has yet been able to show that particular germs are unvaryingly involved in particular IBDs. Some medical scientists and physicians interpret this absence of conclusive evidence as evidence of absence, and pooh-pooh the idea that germs might play any role at all in these diseases. The feeling seems to be that there are too many viable suspects and that they cancel each other out.

The history of microbiology explains some of this feeling. Pasteur and Koch achieved their great breakthroughs by thinking in terms of one germ, one disease, and one treatment fitting together in a unique pattern. Since their day, that approach has continued to pay off in effective insights, treatments, and vaccines. The problem is that IBDs are a family of overlapping and related disease processes, and we can fool ourselves by trying to be overly precise about the distinctions between them. Pneumonia can be fueled by a wide range of infectious germs. Likewise diarrheal diseases have multiple causes and overlapping, sometimes identical patterns. We shouldn't be too surprised if the same thing is true of IBDs. It will probably turn out that a number of infectious germs can lead to serious illnesses like Crohn's disease. The question will likely be to what degree these heretofore ignored suspects do so in varying circumstances, including interactions with other germs.

For example, Italian scientists have reported a possible link between cytomegalovirus (CMV) and some severe cases of IBD. In their study, one-third of a group of patients with so-called refractory colitis, meaning it would not respond to the normal treatment with steroids, harbored CMV, a herpes-family virus; with antiviral treatment, their colitis symptoms subsided.

Evidence is accumulating that at least two other germs fit this pattern: *Mycobacterium avium paratuberculosis* (MAP) and *Bacteroides fragilis* (BF). MAP is an invasive intestinal germ, not part of the normal human flora, that seems to be involved in many cases of Crohn's disease and other IBDs. In the first place, people with Crohn's disease often have antibodies to MAP, indicating past or present exposure to it. Second, antibiotics that knock out MAP bring remission in a large number of cases of Crohn's disease. Third, MAP will trigger Crohn's disease in cattle (called Johne's disease in cattle). Finally, the bacterium has been extracted from the tissues of people and animals with the illness. Recent research has identified MAP in the breast milk of women with Crohn's disease. It survives in milk from infected cows even after pasteurization, which may explain some of its transmission to human beings.

For its part, BF is a normal constituent of the human gut flora; in fact, it is the predominant bacterium in human feces. But some strains produce a potent enterotoxin, and they are designated as enterotoxigenic B. fragilis (ETBF). Direct evidence confirms the presence of ETBF in IBD patients with an active disease process, from thirteen to thirty percent of the entire IBD population. IBD patients whose disease is in a dormant state do not harbor ETBF, indicating that the germ is an important trigger for flare-ups, if not for the disease per se.

Again, I do not think it should surprise us if we cannot yet track the role of specific germs in IBDs with comforting consistency. The digestive process is a complex dynamic system whose elements can interact in a baffling variety of ways. There are some five hundred different kinds of resident germs in the digestive tract, and an uncertain number of transient germs. For two hundred of them we still have only indirect evidence, obtained with the latest techniques of genetic and molecular engineering. We cannot yet isolate these germs as separate microorganisms and culture them in vitro in the laboratory. We've discovered their existence, but we can't yet see them clearly.

My own view is that IBDs will remain problematic until we incorporate holistic thinking into research and treatment. We need to take into account both genetic abnormalities and predispositions on the one hand, and quantitative and qualitative changes in the ecology of the gut flora on the other. In connection with ulcers, I mentioned that a likely genetic factor in that disease, which could also affect IBDs, might be a slower than normal digestive process, which would give troublesome germs a better opportunity to establish themselves where they're not wanted. Or there might be slight abnormalities that raise or lower production of mucus or essential enzymes to a point where digestion becomes unstable. Given the fact that not all of the people exposed to IBD-implicated germs actually develop IBDs, subtle genetic factors like these probably play an important role in determining who gets sick and who doesn't.

Quantitative changes in the gut flora can often trigger an inflammatory process. For example, bacterial overgrowth may occur if undigested food in the lower portion of the small intestine, which is mainly supposed to be an arena for the body's pancreatic and other enzymes, attracts bacteria in the normal flora to creep up from the large intestine. The metabolic by-products of these bacteria can easily irritate the delicate environment of the small intestine. Or a qualitative change in the composition of the gut flora may occur when an invasive germ like MAP or ETBF manages to get a foothold where it didn't have one before. Similarly, a person may pick up a toxin-producing strain of a germ that is usually beneficial, like *Shigella*-toxin-bearing *E. coli* 0157:H7. In this connection, immunologic responses may also have an impact. For example, Th-1 cytokines, which are proinflammatory, can be stimulated out of proportion to their opposite numbers, Th-2 cytokines, which are responsible for activating the production of antibodies. Over time a slight but persistent genetic tendency, such as slower digestion, may facilitate both quantitative and qualitative changes in the ecology of the gut flora. Antibiotics or certain foods may also disturb that ecology.

To understand the role of diet in IBDs, we need to appreciate the normal balance of digestive action between the small intestine and the large intestine. The small intestine receives food from the stomach, breaks it down with the body's own enzymes, and extracts nutrients for absorption into the bloodstream. Some foods are not broken down by the small intestine's enzymes, however, but are passed along to the large intestine to be processed by the bacteria in the normal gut flora.

The small intestine can break down proteins, carbohydrates, and most sugars. But beans, broccoli, cabbage, and other leafy green foods contain a complex sugar called xylose, which normal human enzymes cannot decompose and which makes even a healthy person gassy and flatulent. If you soak these foods in water for three hours, however, draining the water after each hour, the xylose will leach out and almost no gas will be produced. Likewise many people develop an intolerance for the lactose in milk products, because the enzyme needed to decompose it is diminished, or down regulated, in adulthood. A single glass of milk can produce a whole pint of gas in a person who is lactose intolerant. Sugar-free processed foods often contain sorbitol, another substance that the small intestine can't handle. If you look at the fine print on the package, you will see a warning that it can cause diarrhea and flatulence.

Foods that the small intestine's enzymes cannot process pass undigested into the large intestine along with other ingested matter, mucus, and cellular debris. The large intestine is a sort of tubular vat in which the bacteria of the normal flora break down organic matter and ferment it. Ideally the large intestine should have to process only small amounts of undigested food and debris. Its main jobs are to absorb water; take up sodium and other ions to maintain the body's electrolyte balance; make vitamin K, which is absorbed in the colon, the lower part of the large intestine; and form and excrete wastes. The large intestine processes about ten liters of water a day. One and a half liters or so are contained in what we eat and drink; the remaining eight and a half liters or so are produced by the secretions of the digestive glands. The large intestine absorbs ninety-five percent of all this water, keeping the body hydrated and preventing diarrhea. The other five percent is used to form feces, which are about seventy-five percent water and twenty-five percent solid matter. About half of the solid matter in feces is made up of intestinal bacteria; the other half comprises undigested bits of food, mucus, and cellular debris.

Things can easily go awry in this two-stage process even for healthy people, as most of us experience from time to time. For people with IBDs, things can quickly go from bad to worse. Although their small intestines may do well with simple carbohydrates and sugars like those in honey and ripe bananas, people suffering from IBDs may be unable to handle complex-carbohydrate foods like pasta, potatoes, and bread. As I mentioned, undigested complex carbohydrates in the lower portion of the small intestine can attract bacteria from the large intestine. Bacterial growth in the small intestine tends to destroy pancreatic enzymes, further retarding digestion and absorption of complex sugars and carbohydrates and making them available for fermentation in the large intestine. This leads to more gassiness, flatulence, and loose stools. Diets that are rich in complex carbohydrates will also favor the development of certain tapeworms and intestinal protozoa, such as amoeba, whereas a high-protein diet will work against the growth of such parasites. And people who tend to be constipated, because their diets contain too little fiber and roughage, may harbor bacteria in high enough concentrations that their metabolic by-products become carcinogenic as they accumulate and are acted upon by still other bacteria. Dr. Victor Bokkenheuser, long associated with St. Luke's-Roosevelt Hospital in New York City, was one of the first scientists to report on this phenomenon, which is a consequence of what is known as fecal stasis. Or as my grandmother liked to say, "You are what you eat."

With all these issues in mind, let's look at a hypothetical case of a person diagnosed with Crohn's disease and consider different treatments. The standard medical protocol for a serious case would be to focus on the main symptom of the disease, inflammation, and fight it aggressively with steroids. These drugs can indeed reduce the inflammation and provide some relief, but they will not have any impact on its underlying causes. In addition, the benefits of steroid treatment come with nasty side effects that can create their own health problems.

Suppose instead that treatment followed a holistic approach. A logical first step would be to assemble a picture of the numbers and kinds of bacteria in the digestive tract by analyzing the feces. A good microbiologist can tell as much from a person's feces as Sherlock Holmes could from the ash of a suspect's cigar. With this information, a physician could then prescribe an antibiotic cocktail, if appropriate, to reduce any overgrowth of normal flora or knock out invasive germs like MAP and ETBF. In tandem with this, the diet could be adjusted to lessen troublesome foods and add roughage that will stimulate proper bowel movements. (High-fiber diets decrease the transit time of food in the gut by fifty percent compared with low-fiber diets.) Probiotic therapy, adding beneficial germs to the digestive tract, could be the finishing touch. For example, some strains of lactobacilli, which are found in active yogurt cultures or can be bought over the counter in capsules, have a remarkable ability to compete with anaerobic bacteria in the intestines and keep them in check.

Ultimately, gene therapy may become the treatment of choice for inflammatory bowel diseases such as Crohn's disease. In the May 2001 issue of the journal *Nature,* two independent research teams, one at the University of Michigan and the other at Robert Debré Hospital in Paris, reported an exciting new development in the understanding of Crohn's disease. About one in a thousand people normally develop the disease. The two teams found that individuals with an inherited flaw in the Nod 2 gene, however, are seventeen times more likely to contract it. The Nod 2 gene normally synthesizes a protein that helps the immune system distinguish between friendly natural flora and invading bacteria. The flawed gene causes the immune system to attack the body's natural flora, triggering an inflammatory response that can eventually cause the ulceration and erosion of the intestines seen in Crohn's disease. Altering flawed Nod 2 genes could thus conceivably cure many, if not all, cases of this debilitating inflammatory bowel disease.

Until such possibilities become reality, however, many people with IBDs will continue to benefit from multipronged treatment. A great many more will probably continue to suffer needlessly because of entrenched, one-dimensional thinking. Large segments of the medical and scientific community persist in segregating inflammatory bowel conditions from one another and treating and analyzing them in isolation. But this is changing as research continues to build a more sophisticated and detailed picture of how IBDs and other diseases start and progess, especially in terms of the surprising ways in which infectious germs interact with the body. One of the most startling and intriguing of these surprises is the discovery of a connection between Coxsackie virus and a form of diabetes.

Molecular Mimicry

Coxsackie viruses, named for the town of Coxsackie, New York, where they were first isolated, produce a central-nervous-system disease that causes muscular atrophy somewhat like polio, but without paralysis. Scientists have discovered that Coxsackie virus mimics certain cells of the pancreas. Analysis of the gene sequences of a protein on Coxsackie viruses and one on pancreatic cells shows that they are remarkably similar. This suggests that in people infected with Coxsackie virus, the immune system can send antibodies that will attack both the virus and the pancreatic cells it resembles. Over time so many pancreatic cells may be destroyed that insulin-dependent-diabetes mellitus (IDDM) can ensue. This is another example of the cross-reactivity, or cross-antigenicity, that I touched on in Chapter 8.

The theory that Coxsackie virus can trigger an autoimmune response in which the body attacks itself has been tested in experiments with laboratory animals. When animals are infected with Coxsackie B virus, they gradually develop the glucose intolerance that is characteristic of IDDM. Although the autoimmune mechanism remains unclear, additional evidence for it comes from the fact that several outbreaks of IDDM have followed outbreaks of Coxsackie virus infection.

Coxsackie viruses are common and pass easily from one person to another, especially in families. Adults tend to have antibodies for many different Coxsackie viruses that they are exposed to over the course of their lives. This means that children will likely face the greatest risk of being made seriously ill because of an infection with Coxsackie virus or of developing IDDM afterward. The good news is that if research confirms that Coxsackie viruses have a causative role in IDDM in human beings, then there is a strong probability that a preventive vaccine can eventually be developed.

Bacterial Invaders in Our Cells

The ability of infectious germs to make themselves at home among human cells is extraordinary. Yet an even more extraordinary fact of life has been awaiting recognition ever since scientists first learned to look at cells and map their structure. That is that every human cell contains vestiges of bacteria. Indeed, it might be fairer and more accurate to say that all living cells are bacterial in one way or another. The key to this riddle is that there are only two types of cells, prokaryotes and eukaryotes. Bacteria are prokaryotic. The cells of all other living things, from protozoa to plants, primates, and pachyderms, are eukaryotic.

The name eukaryote comes from Greek words meaning "having nuts." The "nuts" in a eukaryotic cell are its own well-defined nucleus and organelles, internal membrane-bound structures that perform specific essential functions for the cell. In plant cells, organelles called chloroplasts are responsible for photosynthesis. In all eukaryotic cells, organelles known as mitochondria act as engines that provide chemical energy for cellular needs. A number of additional organelles serve reproductive, metabolic, or other functions.

Prokaryotes have no internal membrane-bound structures. The "pro" part of the name signifies priority in time, reflecting the fact that bacteria evolved on Earth tens of millions of years before cells with internal structures. In light of these facts, it has been proposed that the mitochondria and other organelles inside eukaryotic cells are actually parasitized bacteria. Four sorts of evidence support this theory. To begin with, physically, biologically, and structurally eukaryotic organelles look just like bacteria. Second, organelles have the ability to make their own proteins, and to reproduce by division, independently of the cell surrounding them, just like bacteria. Third, there are many counterpart examples of bacteria invading and parasitizing eukaryotic cells. For example, rickettsial bacteria like those that cause Rocky Mountain spotted fever are obligate parasites that can survive and divide only inside eukaryotic cells. Finally, extensive DNA research, which was not possible before the invention of the most sophisticated molecular-engineering techniques, indicates that all mitochondrial DNA trace their origin to a single ancestral protomitochondrial genome. This suggests that the mitochondrial organelle was present at the birth of the eukaryotic cell.

In this connection, it is fascinating to note that the Human Genome Project, funded by the National Institutes of Health and the Wellcome Trust, and Celera Genomics, the commercial venture competing with it, both announced in early 2001 that the human genome does not contain a great many more genes than the genomes of other creatures. We human beings are closer than we thought not only to creatures like our chimpanzee cousins, but also to much "lower" life-forms. Instead of the hundred thousand or more human genes that many scientists expected the genome projects to find, there are about thirty thousand, only a third more than a simple roundworm can boast. Moreover, 230 genes in the human genome are identical with genes in bacteria.

So perhaps it should be no surprise that the "bacterial" or "prokaryotic" organelles live in mutualistic harmony with the eukaryotic cells that contain them. The eukaryotes provide room and board, so to speak, and the organelles earn their keep by doing essential maintenance, supplying energy, and so on. Because prokaryotes had aeons in which to evolve before eukaryotes appeared on the scene, they had time to become supremely efficient, making them perfect candidates for incorporation into more complex living organisms.

The pivotal evolutionary question then becomes, did prokaryotic bacteria invade the eukaryotic cells of other life-forms after they evolved, or did eukaryotic cells develop around prokaryotic bacteria in the first place? In other words, are prokaryotes secret agents that have burrowed inside other creatures, or are they conscript soldiers that "higher" life-forms have drafted to fill their needs? A third possibility is that bacteria invaded each other, and that over millions of years of evolution this eventually gave rise to eukaryotic cell types.

Settling this chicken-and-egg question and the others raised in this chapter will doubtless keep science busy for a long time to come. But humanity has crossed a Rubicon in learning to frame questions about the origin of life in such a fundamental way. As we continue to discover the secret agents that cause many of the most feared diseases, we must recognize the extent to which all living creatures depend intimately on the "lowliest" germs. Our new knowledge brings with it an increased responsibility to recognize how human action may adversely affect the great diversity of life on Earth. But if we follow Hippocrates' dictum, "First, do no harm," we can use our one advantage over all other known life-forms—a keen mind capable of both great understanding and remarkable creativity—to maintain the delicate balance of our planet's ecosystem.

11

The Germ Revolt

Modern medicine is running out of magic bullets. Multi-drug-resistant tuberculosis is spreading worldwide, as we saw in Chapter 8, and has already appeared in seventeen American states. Toxin-producing strains of *Staphylococcus aureus* and *Escherichia coli* have become resistant to a wide spectrum of antibiotics, and some *S. aureus* germs are demonstrating so-called intermediate, or partial, resistance to what has until recently been the antibiotic of last resort, vancomycin. In regard to health, VISA (vancomycin-intermediate-resistant *S. aureus*) is everywhere we don't want it to be. In the ranks of drug-resistant germs, it stands shoulder to shoulder with the thirty percent of *Streptococcus pneumoniae* strains in the United States that are partially or fully resistant to penicillin. Some strains of enterococcus that cause urinary-tract and heart-valve infections resist all so-called first-line antibiotics, forcing reliance on a dwindling list of substitutes. A drug-resistant strain of *Salmonella* recently popped up in Denmark, causing an outbreak of illness that struck twenty-seven people and killed two of them. Likewise, in Kenya strains of typhoid-fever germs *(Salmonella typhi)* will no longer succumb to treatment with an antibiotic called ciprofloxacin. And the list goes on and on. If any of these antibiotic-resistant strains of infectious germs has not yet made it to the United States, that is surely only a matter of luck, and temporary luck at that. Global travel and trade can bring them to our doorstep at any time.

Wonder Drugs and Super Bugs

The great German scientist Paul Ehrlich fired the first "magic bullet," as he dubbed it, in 1909–10, when he and his Japanese assistant, Sahachiro Hata, identified an arsenic compound, the 606th one they tried, that was effective against *Treponema pallidum,* the spirochete bacterium that causes syphilis. Ehrlich, then director of the Royal Prussian Institute for Experimental Therapy in Frankfurt, Germany, had begun his research career by testing industrial dyes for their action on cells. Particular dyes stained some cells but not others. This finding paralleled two other discoveries of late-nineteenth-century medical research: that particular bacteria and fungi would kill, or inhibit the growth of, some germs but not others, and that the body's immune system could distinguish between "self" and "nonself."

In 1877, Louis Pasteur observed that certain soil microbes killed anthrax bacilli, and he and others subsequently made similar observations about other substances and germs. Although they didn't realize it at the time, these investigators were observing an antibiotic phenomenon: When the soil microbes found themselves competing in the same environment with the anthrax bacilli, they released a substance to fight the rival germs. Despite his inability to explain this phenomenon, Pasteur astutely suspected that it could have important implications for the development of new medicines. But he was fully occupied with research into vaccines and could not take time to explore the possibility in any depth.

In 1882 Élie Metchnikoff, a Russian scientist who later became Pasteur's colleague at the Pasteur Institute in Paris, began to uncover the secrets of cellular immunity while studying starfish larvae. Metchnikoff noticed that the larvae contained mobile cells whose purpose he tested by introducing small thorns from a tangerine tree. Within twenty-four hours, the mobile cells had surrounded the thorns and were trying to engulf them. This was the first recorded instance of what Metchnikoff came to call phagocytosis, the process, as we saw in Chapter 4, whereby immune-system cells engulf intruder germs and other foreign substances.

Ehrlich brought these two strands of research together with his own work on dyes in an effort to find chemicals or other substances with selective toxicity, "magic bullets" that would fly straight to the disease-causing germs in a person's body without harming the body's own cells. Even before his syphilis breakthrough, the results were dramatic enough to win Ehrlich the 1908 Nobel Prize for Physiology or Medicine, which he and Metchnikoff shared for their separate contributions to the understanding of immunity. The arsenic cure for syphilis, appealingly named Salvarsan 606, brought him even greater fame and made him a household name long after he died, of a stroke, in 1915. Edward G. Robinson even starred in a movie about him, *Dr. Ehrlich's Magic Bullet,* that was released in 1940. Ironically, at about that same time the Nazi government in Germany was trying to obliterate Ehrlich's memory because he was a Jew. The street containing the Royal Prussian Institute for Experimental Therapy had been named Paul Ehrlichstrasse in his honor; the Nazis tore down the signs and renamed the street. Fortunately they could not erase his name from the world's history books, and Paul Ehrlich will continue to be remembered as the father of chemotherapy.

Salvarsan saved a great many lives, but it lacked some of the magic Ehrlich had envisioned. It had two main drawbacks. First, effective therapy involved a series of painful injections over a protracted period. Second, Salvarsan's selective toxicity could be indiscriminate enough to harm patients. But the quest begun by Ehrlich went on, inspired by the prospect of adding more magic bullets to medicine's armory. All the magic that medical science could dream of finally seemed to arrive with penicillin, discovered by Alexander Fleming in 1928. Fleming, a Scottish bacteriologist at St. Mary's Hospital in London, returned from a vacation to find that a fungal mold had destroyed some staphylococcus colonies he had been growing in a petri dish. Other investigators had made similar observations before, but Fleming took things a step further. He isolated penicillin from the mold and tried to use it therapeutically. Unfortunately the strain of penicillin Fleming had to work with was not potent enough to have much effect, especially in light of the fact that it was hard to produce in any quantity.

The story might have ended there, except that ten years later two scientists at Oxford University, Howard Florey, an Australian pathologist, and Ernst Chain, a German biochemist who had fled the Nazis, decided to take another look at penicillin. The difficulty of growing penicillin in fungal molds was such that only one part in two million was pure penicillin. With input from Norman Heatley, another biochemist, production of penicillin improved to the point where, in May 1940, it could be tested successfully on mice. Emboldened by this victory, the Florey-Chain team then tried to use it on a human patient, a policeman who had scratched himself while pruning a rosebush and then come down with staphylococcal blood poisoning. Nowadays such a scenario can hardly fail to elicit a smile. But before antibiotics, it was no joke; the policeman was on the brink of death when penicillin was administered. The new drug was in such short supply that the policeman's urine was collected so that as much penicillin as possible could be extracted and recycled. The results of this first human trial of penicillin were both tantalizing and tragic. For a few days the policeman made an astonishing recovery. But there simply wasn't enough penicillin to sustain the process, and on the fourth day he died. In medical lore he is remembered, using the British slang term for a policeman, as "the Oxford bobby."

As World War II raged on, the potential value of penicillin became ever more tantalizing. Eventually, Florey and Chain took the production problem to the United States government's northern Regional Research Laboratory in Peoria, Illinois. There an American-British team figured out how to brew penicillin in beer vats, and before long both American and British pharmaceutical companies were producing large quantities of high-grade penicillin. With ample supplies there was no longer any question about the results. Penicillin saved soldiers' lives on D day and in every subsequent battle of the war, fully meriting the Nobel Prize that Fleming, Florey, and Chain shared in 1945. Since then penicillin has saved countless lives from the ravages of infectious disease. Among the main ailments it fights effectively is Ehrlich's great antagonist, syphilis.

Survival of the Fittest Germ

Success breeds success, and the development of antibiotics quickly accelerated. In 1944, Selman Waksman discovered that streptomycin was effective against tuberculosis, and in the 1950s and 1960s pharmaceutical companies developed and marketed a broad array of antibiotic wonder drugs, from cephalosporins to tetracycline. Yet there were problems with antibiotic-resistant germs right from the start. Tuberculosis bacilli quickly developed resistance to streptomycin, for example, and physicians were forced to prescribe antibiotics in combination to halt the disease, as they still must do. In the 1950s about half of all *S. aureus* strains were already resistant to penicillin. So long as antibiotics research and development remained in full swing, however, medical science seemed more than able to keep pace with the emergence of resistant strains of infectious germs. But in the late 1970s and early 1980s, with more than a hundred powerful antibiotics on the shelf to be used singly or in combination as needed, the big pharmaceutical companies downplayed antibiotics R&D and began to look elsewhere for new products to bring to market. As often happens in human life, overconfidence and complacency set the stage for a dramatic turning of the tables. Over the last two decades antibiotic resistance has spread far and wide, a deadly germ revolt that we have only recently begun to recognize and respond to effectively.

To understand how antibiotic resistance arises in germs, we need to understand the basics of how antibiotics work. The word itself holds an important clue. It literally means "anti-life." Antibiotic therapy pits the selective toxicity of a good germ, or a chemical substance derived from it, against a bad germ in the environment of a person's body. A no-holds-barred, Darwinian struggle ensues, which determines whether or not health is restored.

Antibiotics can attack or inhibit the growth of germ cells in five ways. They can interfere with cell-wall synthesis, as penicillin and vancomycin do, causing infectious germ cells to burst and die. Or like erythromycin and tetracycline, they can disrupt cellular processes such as protein synthesis, preventing further growth and making it easier for the immune system to knock out the unwanted germs. A class of antibiotics known as polymyxins can undermine infectious germs' ability to take in nutrients and expel toxins and wastes through the cell envelope, and this loss of cell-membrane permeability will in time cause cellular death. Antibiotics like ciprofloxacin and levofloxacin can keep germs from reproducing by interfering with DNA synthesis. Finally, some antibiotics can compete with nutrients to be taken up by germs. For example, certain bacteria need to take up a vitamin substance called para-aminobenzoic acid (PABA) in order to grow. Sulfa drugs so closely resemble PABA molecules that they trick the bacteria into picking them up instead. This also inhibits bacterial growth and buys the immune system extra time in which to do its work.

Despite Ehrlich's vision of "magic bullets," antibiotics are not really one-to-one agents. They affect groups of similarly constituted bacteria, fungi, and protozoa, but not viruses. All such groups contain resistant strains, simply by virtue of the hit-and-miss mutations that occur on average about once in every million cell divisions. The more antibiotics are used in a particular environment—a person's body, a household, a hospital, or the world at large—the more they will upset the natural competitive balance among germs in favor of resistant strains. For example, if a teenager takes antibiotics regularly for acne, the other members of the household will soon come to harbor high concentrations of antibiotic-resistant skin germs simply by ordinary person-to-person contact. Spread throughout the germ world are resistance traits to counter every strategy that antibiotics employ. Some germs can produce enzymes that disable antibiotics. For example, a number of germs can produce an enzyme called penicillinase, which breaks down penicillin and renders it ineffective. Others can change their binding proteins so that antibiotics cannot identify and latch on to them. Still other germs have evolved pumps that eject attacking antibiotics before they can reach their targets inside the germs' cells. Tetracycline is particularly vulnerable to this phenomenon, which is known as efflux.

Most resistance genes present in bacteria reside on plasmids, tiny loops of DNA that are not attached to the bacterial chromosome. Resistance genes can be passed along in three ways. In "conjugation," as scientists refer to it, one germ extends a sexual pilus to another germ to give it a plasmid. In the process called "transformation," germs pick up scattered DNA segments after a germ cell has died and spilled its contents into the environment. And in "transduction," a bacterial virus extracts genetic elements from one bacterial cell and injects them into another. Resistance genes often ride on DNA units called transposons, which as their name suggests can easily shift from one DNA molecule to another. Many bacteria have specialized transposons, called integrons, that attract new genes like a magnet.

An additional complication is that antibiotics will kill or inhibit bystander germs that resemble the disease-causing germs at which they are aimed. Many of us experience this sort of collateral damage when antibiotics, taken for a problem elsewhere in the body, disrupt the normal gut flora and trigger diarrhea. Antibiotic action on bystander germs also enhances the spread of resistance genes. First, it reduces the number of natural competitors for living space and resources that a resistant infectious germ would otherwise have to confront. Second, it favors the survival of resistant strains of innocuous germs as well as dangerous ones. And a harmless germ can acquire and pass on resistance genes as easily as a harmful one can. Gene exchange between two germs can take less than an hour. The forty-to-sixty-hour cycle of human digestion leaves ample time for this to occur in the human intestinal tract, making it a frequent arena for horizontal transfers of genetic material between germs.

Finally, resistance genes can be picked up and shared far from their source, once they have been introduced into the food chain. The heavy use of antibiotics in agriculture and the livestock industry insures that this scenario is played out over and over again around the world. To return to an example I mentioned at the beginning of the chapter, the antibiotic-resistant *Salmonella* that killed two people in Denmark was traced to antibiotics in pig feed. Likewise, in Holland strains of *Enterococcus faecium,* a common human intestinal germ, apparently acquired antibiotic resistance because of the eating of turkeys whose feed contained antibiotics. Although these germs will not themselves make the people who harbor them sick, they stand ready to pass their resistance traits to any infectious germs that manage to find their way into the same environment.

All these varied genetic transfers would probably pose no very great problem, however, if we used antibiotics only when health needs specifically required them and in strict accordance with sound medical protocols. But we don't. In the first place, people often stop taking antibiotics when their symptoms clear, instead of finishing the prescription. This allows antibiotic-resistant germs to survive to cause illness with renewed force sometime later. People also very often share antibiotics with one another, spreading them around enough to favor resistant strains of harmful and harmless germs alike in two or more people rather than just one.

Then there is the question of overprescription. Every year in the United States 150 million prescriptions, sixty percent of the total of all prescriptions, are written for antibiotics. Fifty million of them are unnecessary and, in societal terms, medically dangerous. They are given for respiratory infections, ear infections, and a host of other ailments that are caused by viruses, rather than bacteria, fungi, or protozoa. Antibiotics are useless against viruses, but all too often people suffering from a cold or flu insist on them, and all too often harried physicians go along rather than argue with misguided but determined patients. The situation is even worse outside North America and Western Europe. In the Third World, antibiotics are generally available over the counter without a prescription. Residents and travelers alike can self-prescribe these drugs, turning themselves into vessels of antibiotic-resistant germs when there may be no good reason for doing so.

Worldwide, antibiotic use in agriculture is soaring. Not only are antibiotics mixed into feed for cattle, pigs, and poultry to promote speedy growth and bring healthier, more profitable animals to market, they are also sprayed onto fruit trees and other harvest plants. The Union of Concerned Scientists estimates that the American agricultural industry puts about twenty-five million pounds of antibiotics a year, seventy percent of the country's total antibiotics production, into livestock feed and other nontherapeutic uses. Therapeutic uses, when sick animals need treatment, consume additional antibiotics. Yet according to the Union of Concerned Scientists, only about three million pounds of antibiotics are used for treating disease in people. The agricultural industry claims very different numbers, of course, but we should ask whether the benefits of any large-scale use of antibiotics in agriculture justifies the medical costs. It is conservatively estimated that antibiotic resistance costs U.S. taxpayers about $5 billion a year. The cost in lost lives and prolonged suffering may be incalculable;

it is certainly rising year by year as antibiotics use increases and antibiotic resistance spreads.

Finally, we need to acknowledge the extent to which we make antibiotics use necessary through carelessness and neglect. As I've already mentioned in different contexts above, proper hand washing and food hygiene would dramatically reduce the rate of infectious disease and thus the rate of antibiotic prescriptions. Following the commonsense food hygiene **Protective Response Strategies** outlined in Chapter 7 could eliminate half the hundred million food-borne illnesses that occur in the United States every year. Eighty percent of all infections are transmitted by touch, and hand washing could eliminate the vast majority of them, as well. Proper hand washing in hospitals would have a welcome effect on the appalling rate of nosocomial infections that I discussed in Chapter 2. Remember that these nosocomial infections directly cause the deaths of thirty thousand people a year, and materially contribute to the deaths of seventy thousand more. Breast cancer, by contrast, killed about forty-four thousand people in the United States in 1998, and AIDS killed about seventeen thousand. According to the CDC, 2.5 million people a year pick up an infection inside American hospitals. The total cost of treating these nosocomial infections is estimated to be $4.5 billion a year.

Antibiotics have saved millions of lives over the last half century, and they can save many more in the decades to come. But in one sense the world's antibiotic use has been a fifty-year experiment in self-sabotage. Without our realizing it, the selective toxicity of antibiotics has bred more and more dangerous germs. Wonder drugs have produced super bugs. To lessen the extent to which this keeps happening, and insure that antibiotics remain potent weapons against disease, a multifaceted **Protective Response Strategy** is needed.

- **Understand what antibiotics can and cannot do. Never pressure a physician to prescribe them inappropriately.**

- **Be a good patient. If you have a condition that requires treatment with antibiotics, follow your doctor's instructions** carefully. Take the full prescription as directed, and never share or hoard antibiotics.

- Help to get antibiotics out of the food chain by buying meat and eggs labeled as coming from animals whose feed contained no antibiotics. If enough consumers express their views in this way, it will powerfully motivate the agricultural industry to find alternative ways of enhancing the growth of animals and other foods. You can also let your elected representatives know that you are concerned about this public-health issue, and that you want to see the agricultural industry reduce its reliance on antibiotics.

- Practice good personal hygiene, and wash your hands frequently as suggested in the Protective Response Strategy in Chapter 2. Likewise, practice good food hygiene as outlined in the Protective Response Strategy in Chapter 7. Both efforts will reduce antibiotic resistance by reducing the rate of infections and thus the need for antibiotics to be prescribed in the first place.

- Use germicides and appropriate germicide-treated products. These, too, can help to reduce the rates of infection and antibiotics use. An added benefit of germicides is that they often have a residual effect. For example, hand washing with liquid soaps containing germicides will leave an unnoticeable residue on the hands that will continue to fight germs for hours. Not all germicidal products are worthwhile, however, and there is a simmering controversy in the medical profession about any household use of germicides. To choose sensibly from the growing list of such products, consumers need to educate themselves about that controversy.

The Germicide Controversy

Although a medical consensus grants that germicides have a place in hospitals, clinics, and so on, some scientists argue that widespread use of germicides by the general public contributes to antibiotic resistance. They claim that normal detergent, soap, and water, with plenty of elbow grease, should be more than enough for personal and household cleanliness.

Along with many other scientists, I disagree. In my view germicides are not part of the problem of antibiotic resistance, they are part of the solution. In the first place, germicides act in a way very different from antibiotics. They do not attack some varieties or strains of germs, while sparing others, as antibiotics do. Instead they inhibit all germs, including viruses, to varying degrees in varying concentrations. Because they don't play favorites, germicides don't upset the natural competitive balance among germs. They inhibit antibiotic-resistant and nonresistant germs alike.

Germicides entered regular medical practice in the late 1860s, when Joseph Lister began campaigning to encourage cleaning both wounds and operating rooms with carbolic acid. Hospitals have relied on a growing range of germicides ever since, and they have been incorporated into a large number of household products from toothpastes, mouthwashes (Listerine is an early example of this), deodorants, and soaps to cutting boards, carpeting, and toys. Yet not a single instance of antibiotic resistance can be traced to real-world use of germicides. During the same period, germicides have proven to be remarkably safe. Most germicidal products contain a mix of substances, but the commonest, most frequently studied, and most controversial germicide is triclosan. Numerous tests—for toxicity, allergy, carcinogenicity, photosensitivity, and mutation—have failed to turn up any damaging side effects from triclosan.

One recent study has suggested that triclosan may indeed contribute to antibiotic resistance. But this study strikes me as seriously flawed. At best, it demonstrates that some germs have a certain tolerance for germicides, so that more germicide is needed to inhibit their growth. To begin with, the study used laboratory-manipulated strains of E. coli that are not representative of what would be found in nature. It did not test triclosan's action on very differently structured bacteria such as *S. aureus,* and it did not test the germicide's effect on viruses at all. The study also failed to address the fact that most over-the-counter germicides contain multiple antibacterial substances, ignoring the need to address the synergistic effects that germicides can have in combination with one another.

In my own research I have tested triclosan's action on hundreds of very antibiotic-resistant and very antibiotic-susceptible bacteria. I have found that there is a slight difference in triclosan's effect on the two groups: The germicide actually kills or inhibits the growth of antibiotic-resistant bacteria a little bit better than it kills or inhibits antibiotic-susceptible bacteria. This may be because antibiotic resistance consumes resources that would otherwise bolster a germ's general hardiness. In any case, the results of hundreds of experiments strongly support the view that germicides can help us to combat antibiotic resistance, rather than contribute to the growth of the problem.

The distrust of germicides in some quarters eerily echoes the mid-nineteenth-century medical establishment's treatment of Ignaz Semmelweis. I mentioned in Chapter 2 how Semmelweis dramatically reduced deaths from childbed fever in maternity wards in Vienna and Budapest by insisting that caregivers wash their hands before examining and treating patients. Yet Semmelweis's colleagues and superiors refused to believe the evidence in front of them and actively discouraged the hand washing that was saving women's lives. It is interesting to note in this regard that Semmelweis found that plain soap and water didn't do a good enough job on their own. Before any significant lifesaving results could be achieved, he had to insist that medical students and other caregivers wash with chlorinated lime. It is not that soap and water cannot remove germs from the hands. They can, but only if they are properly used.

The CDC recommends washing hands for fifteen seconds with plenty of soap and water, taking care to wash between the fingers and under the nail beds and to work soap into the creases around the knuckles, all places where germs can lurk, then rinsing thoroughly and repeating the whole process. The International Food Safety Council advises washing hands for a full twenty seconds, with a twenty-second repeat. But in the real world, this is not how people wash their hands, if they wash them at all. As we saw in Chapter 2, more than fifty percent of men and women alike fail to wash their hands after using toilet facilities, although women are a little more careful than men. When people do wash, they usually simply run water over their hands or fingers for a few seconds, perhaps with a useless dab of soap and perhaps without. Whether judged on time or technique, doctors, nurses, and other hospital staff all too often do no better. As a result most hospitals try to insure that hand washing on their premises is done with germicides, rather than plain soap and water.

In short, it seems to me that germicides can be a valuable support to good health, and a helpful ally against antibiotic resistance, especially given how hectic and pressured modern life can be. If we could all take the time to wash properly on every occasion, soap and water would do just fine. But most of us spend our days rushing from one thing to the next. We also may not always have access to soap and water when we need to wash our hands. I carry a small bottle of liquid waterless germicide and find it very useful in a pinch. By the same token, it is hard to see a downside to using germicidal lotions, soaps, and other personal care products in the home. They save time and have no side effects, and the existing evidence indicates that they will not contribute to increased antibiotic resistance.

In Chapter 7, I said that a germicide-impregnated cutting board would be an asset in any kitchen. Even a well-cleaned board will likely harbor germs in the fissures that a knife cuts in its surface. A germicide works to eliminate those germs in between uses. Utensils with germicide-impregnated handles strike me as worthless, in contrast. Consider a pizza cutter with a germicide-impregnated handle. Most of the potential for germs to collect and grow in bits of sticky cheese and other pizza ingredients will be on the blade of the cutter, not the handle. And if you wash the blade, you will surely wash the handle, too, won't you? Products like this only succeed in giving germicides a bad name.

Toys made from plastics and resins that have been infused with germicides are another good idea. Like cutting boards, such toys often suffer nicks and scratches that provide shelter in which germs can flourish. The constant degerming action of antibacterial substances can significantly reduce the number of germs present on the toys even while their owners are tucked away in bed.

The number of household germicidal products has more than doubled over the last few years. To pick the good ones from the bad ones, consumers must heed the challenge of "buyer beware" and not expect germicides to do more than they can. It should go without saying that germicides are no substitute for conscientious, commonsense cleanliness and food hygiene. But in the face of the threat of rising antibiotic resistance, germicides buy a little extra precious time as we search for new antibiotics and strive to sustain the effectiveness of existing ones. Fortunately there is encouraging news on both counts. Until very recently it looked as if there would soon be no drug left to combat potentially deadly staphylococcal infections. But even though VISA (again, that's vancomycin-intermediate-resistant *Staphylococcus aureus*) is everywhere we don't want it to be, two new antibiotics, synercid and linezolid, have been brought to market by different manufacturers, and they have so far proven very effective against staphylococcal germs.

For their part, old standbys like penicillin still have plenty of fight left in them, at least as far as some deadly germs are concerned. Penicillin remains marvelously effective against *T. pallidum,* the germ that causes syphilis. Dr. Ehrlich's Salvarsan 606 was a magic bullet against syphilis for a season, as it were, but penicillin seems to be a magic bullet for all seasons. Penicillin also continues to do well against the flesh-eating bacterium group A beta strep. Apparently these germs lack some of the genetic resourcefulness that other germs boast.

As the phenomenon of antibiotic resistance shows, the evolutionary adaptability of germs is almost limitless. But so, too, are human creativity and understanding. If we treat germs with the respect they deserve, restraining our use of antibiotics within more reasonable bounds and using all our skills and insights to plumb the mysteries of the germ world, then my bet is that we can hold our own and even achieve new antibiotic breakthroughs, as well as novel, alternative ways to combat pathogenic germs. We'll explore these possibilities in the last chapter of this book, but now let's turn to another urgent problem, the potential for germs to be used as weapons of war and terrorism.

12

Germ Warfare and Terrorism

It hasn't happened—yet. But consider the following scenario.

Terror by Night

December 31st, 11 P.M.: In a deserted hangar at the Old Rhinebeck Aerodome in Rhinebeck, New York, about a hundred miles up the Hudson River Valley from New York City, a man readies a crop-dusting plane for a special New Year's Eve flight. Midway through his preflight checklist he takes a vial from his pocket and swallows his third dose that evening of doxycycline, an antibiotic, as a pro-phylactic measure. He doesn't really fear being infected tonight; his big risk of exposure is past. And ultimately he doesn't care when he goes down in flames. But he is determined to live long enough to enjoy the statement he's about to make.

It's a statement he's been planning for two years now, ever since his dismissal as a civilian laboratory technician from the U.S Army's Medical Research Institute of Infectious Diseases at Fort Detrick, Maryland. His supervisor said he had been "consistently lax in the handling and treatment of hazardous materials." That was a lie. Well, now he'd show that jerk supervisor and everybody else in the lab that he knew what he was doing. Now he'd show the whole world.

Deciding what to do and how to do it had been easy. The army's own antibioterrorism manuals had given him the basic concept, although when and where were strictly his idea. Fortu-nately, he had a little capital to work with. His parents' dinky house in Camden, New Jersey, had been sold the year before, after his mother died, and the money was still sitting in the bank. It was enough to live on, if he was frugal, until he put his plan into effect. After that, money wouldn't matter anymore.

The tough part had been getting his pilot's license and, even more of a problem, getting access to a crop-dusting plane. He earned his license a year after being canned at Fort Detrick. Then he spent months trying to hire on with one of the handful of outfits that did crop dusting in the New York, New Jersey, Connecticut area. For a while it seemed as if he might have to wait another whole year to execute his plan. Then a pilot for Catskills Crop Dusting took off for parts unknown just as the busy season was heating up, and a job opened up for him. It was like a sign from fate that his plan couldn't fail.

One last check of the instrument panel and he's ready for takeoff. He wishes he could fly straight over New York City and the party in Times Square, but that's a little too risky. Besides, the real fireworks will come in a few days, if the germs get a chance to do their stuff. He feels like it's up to him to see that they do.

He flies south to within twenty miles of the city and then circles around to the northeast, where the prevailing winds are coming from that night, exactly as forecast. He turns the plane into the wind, checks his course, and then, with the lights of the city glittering behind him, releases the crop duster's payload. He giggles as two hundred pounds of powdered anthrax begin floating down toward the city. The particles are very small, five microns or less in diameter, so that gravity doesn't drag them down too quickly. By the same token, the nighttime temperature inversion will keep the particles from rising too high in the air. By the time they reach New York City, they should be settling down to just about street level.

It had been a snap to get the anthrax. Four months before his dismissal, a batch of samples had come in from the Middle East, where the soil commonly contains anthrax bacilli. He'd been assigned to work with the samples, isolating the microbes and growing out colonies to check that they were susceptible to ciprofloxacin and doxycycline and to see if any mutant strains had occurred in the wild. He slipped a couple of tablespoons' worth of soil into his pants pocket and took it home to culture anthrax bacilli in petri dishes on his kitchen windowsill. At first he hadn't known what he was going to do with the anthrax. But it was exciting to have, better than any gun. He liked to imagine what it could do if it were released in the air-conditioning vents of a shopping mall or dropped over a St. Patrick's Day Parade, or best of all, scattered over Times Square in New York City on New Year's Eve. The whole world would see a mass murder take place, and only realize it after the fact!

Then they fired him, and that became his mission. With his lab skills it was simple enough to grow the anthrax colonies he needed, subject them to desiccation so that they would sporulate, and then grind the spores up into powder. While doing this work, he made sure to wear protective gear and dose himself with doxycycline or ciprofloxacin, also courtesy of the lab at Fort Detrick. He'd been helping unpack a shipment of supplies when he got the chance to steal the antibiotics, stashing them behind the cabinet and then announcing that the shipment and the packing slip didn't agree. Suspicions about that had probably helped get him fired. But one of the best parts of the whole scheme was all the flak his old bosses were going to catch for being "consistently lax in the handling and treatment of hazardous materials." He was going to get a big kick out of that.

Ciprofloxacin and doxycycline are the primary antibiotics that offer protection against infection with anthrax bacilli, if they are taken in advance of, or very soon after, exposure. The man in the plane doubts that any more than a handful of medical experts and counter-terrorism personnel will be on the right antibiotics on this New Year's Eve. For the thousands of people celebrating in Times Square or out on the streets elsewhere as the anthrax bacilli float down, inhaling them will be an almost certain death sentence. After a one- to six-day incubation period, infected people will begin to feel as if they are coming down with the flu. If any of them are lucky enough to be treated with ciprofloxacin or doxycycline—a remote possibility, he figures—they might pull through. But the odds are that suspicions of anthrax won't be aroused until the next day. Then people will begin to suffer respiratory distress, in many cases with a telltale pain in the middle of the chest called mediastinal involvement. As clusters of people with these symptoms turn up in emergency rooms all over the tristate area, some alert doctor, nurse, or public-health officer may raise an alarm about a possible outbreak of anthrax infection. But by then toxins will be coursing through the victims' bloodstreams, and it will be too late to save their lives. Within about twenty-four hours they will go into shock, and shortly afterward they will die.

The man in the plane can see it all happening in his mind's eye. He can't wait to see the video on CNN. Sooner or later, he knows, the authorities will get on his trail. But that doesn't matter. They could arrest him in midair, for all he cares. . . .

Okay, my plot is full of stretches and coincidences, and I'm obviously no threat to John LeCarré, but I've spun the scenario out at this length in order to highlight some of the most important issues in bioterrorism and biological warfare. Anthrax could indeed be effectively delivered with a crop duster, and if it were—by a disturbed individual or a team of terrorists—it would probably exact a very heavy toll. Congress funded a study which reported that two hundred pounds of powdered anthrax spores floating down on the Washington, D.C., area would kill up to three million people. A cigar box can hold enough anthrax to kill tens of thousands of people.

These figures are not just the result of "what if" projections. The Russians have had some unfortunate real-life experiences that show how easily people can be killed by the release of anthrax bacilli into the air. In 1979, workers at an anthrax plant in Sverdlovsk, in central Russia, unwittingly sprayed anthrax particles into the outside air when they operated a drying machine with a missing filter. About four hundred people, some living three miles away, were made sick, and about ninety people died. A flock of sheep grazing some thirty miles away also died. In 1992, Boris Yeltsin admitted that sixty-six people living downwind from a Russian government lab had been killed by another accidental release of anthrax.

A Big Bang for the Buck

The Russian government, Soviet and post-Soviet, is hardly alone in making biological weapons part of a national arsenal. Biological weapons are attractive to governments and terrorists alike for the same reason: They deliver a big bang for the buck. Compared to nuclear weapons or conventional armaments on ships, tanks, and planes, biological weapons are cheap and easy to make. For any individual or organization with a rudimentary understanding of microbiology and the requisite materials, making biological agents of death in quantity is little more demanding technically than brewing beer. The microorganisms involved are often readily obtainable in nature, like the anthrax bacilli that abound in soils in the Middle East, or can easily be acquired from sources such as a country's pharmaceutical and agricultural industries. Toxic ricin, for example, which strikes at the central nervous system, can be extracted from the same castor beans that are the source of castor oil.

The United States Department of Defense has published a list of the seventeen likeliest biological weapons. They fall into three categories. The first includes deadly bacteria such as anthrax and plague germs. The second comprises viruses, such as those which cause smallpox, encephalitis, and hemorrhagic fevers like Ebola, Lassa, and Rift Valley fever. The last group is made up of toxins that attack the central nervous system, such as botulinum, fungal toxins, and ricin.

	POTENTIAL BIOLOGICAL AGENTS	VACCINE AVAILABLE
1	Anthrax	Yes
2	Botulinum Toxins	Yes
3	Brucellosis	Yes
4	Cholera	Yes
5	Clostridium Perfringens Toxins	No
6	Crimena-Congo Hemorrhagic Fever	Yes
7	Melioidosis	No
8	Plague	Yes
9	Q Fever	Yes
10	Ricin	Yes (experimental)
11	Rift Valley Fever	Yes
12	Saxitoxin	No
13	Smallpox	Yes
14	Staphylococcus Enterotoxin B	Yes (experimental)
15	Trichothecene Mycotoxins	No
16	Tularemia	Yes
17	Venezuelan Equine Encephalitis	Yes

Vaccines are available for many of these biological agents, as you can see above. But in the absence of vaccination programs for the general public, the nonexperimental vaccines have so far been used in the United States only to protect soldiers going overseas, counterterrorism units, and some so-called first responders—the emergency medical, fire, and police personnel who would be dispatched to the site of a bioterrorism attack to try to contain the damage and treat the victims.

Delivering the right treatment to victims would be one of the most troublesome aspects of any act of bioterrorism. The first indication of an attack would likely come some time after the fact, as people began falling ill in an unusual way. But an unusual pattern of illness takes time to emerge. The incubation period that usually intervenes between exposure to an infectious germ and the first appearance of symptoms in infected people will be one source of delay. Another will be the problem of recognizing whether an illness has a natural cause or not, especially if its early stages mimic familiar disorders like the flu. These two factors will handicap medical personnel even if they have immediate access to effective antibiotics and antitoxins. By the time a correct diagnosis is made, it may well be too late to save lives.

A real-life incident at the U.S. Army's Medical Research Institute of Infectious Diseases at Fort Detrick, Maryland, illustrates how difficult diagnosis can be. In March 2000, a thirty-three-year-old civilian microbiologist working in one of Fort Detrick's high-security labs became ill with fever. The fever started at 100.5 degrees Fahrenheit, then rose quickly to 103 degrees Fahrenheit, accompanied by general malaise and some weight loss over the course of a few days. The researcher went to his private physician, who treated him with antibiotics, and he continued to go to work. But he didn't get better. Eventually, he had to be hospitalized, and finally with the help of U.S. Army doctors he was diagnosed with glanders, a disease of horses that is fatal to human beings if untreated and that has been the subject of weapons experiments by the United States, the Soviet Union, and other countries. The scientist had been working with eighteen other researchers under rigorous safety conditions in a tightly sealed laboratory with elaborate filtration systems. He had so much confidence in these safety measures that he didn't consider the possibility that something in the lab had made him sick, especially when none of his coworkers became ill. Luckily for him an accurate diagnosis was reached in time, he was treated properly, and he recovered. But if people who work with biological agents every day can be fooled, think how easy it would be for ordinary emergency-room staff and others to interpret the result of a bioterrorist attack as a simple case of the flu or some other common malady.

As far as terrorists are concerned, the time lag between exposure to a biological agent and the appearance of symptoms may be both a plus and a minus. On the one hand an incubation period of a few days or longer gives terrorists time to attack and escape before an alarm is sounded. On the other hand too long an incubation may diffuse the impact that the terrorists are hoping to achieve in order to bring attention to themselves, their cause, and their grievances.

In this light we can see that anthrax deserves its number-one ranking on the top-seventeen list because of a combination of circumstances. It is highly lethal, relatively fast-acting, and best of all it can be delivered efficiently in aerosol form by sprayer or crop duster, subject of course to the vagaries of weather and terrain. Other biological agents, like the nerve gas sarin, can also be delivered in droplet nuclei that can injure or kill a person if they are inhaled or even if they are brought into contact with bare skin. But they are likely to be more effective in an enclosed space than in open air, limiting the number of potential victims. This does not mean that the results of such an attack would not be devastating. When the radical Japanese religious sect Aum Shinriyko released sarin nerve gas in the Tokyo subway system in 1995, scores of people were killed or injured. The possible spread of germ-warfare agents in the New York City subway system has been tested with *Serratia marcesens,* a relatively innocuous germ that grows as red-colored colonies, which allows their presence to be detected easily. It was found that the suction of trains traveling through the New York City subway tunnels could easily carry germs from one station to another.

Germs that an impatient terrorist might reject as too slow-acting might prove formidable weapons in a war. A highly contagious germ like *Yersinia pestis,* the cause of pneumonic plague, which can be passed from one person to another in sweat and respiratory secretions, could deliver a vicious one-two punch, as people made sick by an act of war infected others before they died. Even the most sophisticated health-care system would surely buckle under the strain of such a modern plague, if one were ever unleashed.

From Poisoned Wells to Designer Plagues

Biological warfare began in the distant past, when the first enemy waterhole was poisoned. For a long time this was probably the limit of biowarfare. New biological weapons or tactics depended on advances in knowledge, or lucky guesses, about how infectious diseases spread. The first recorded instance of more extensive biowarfare occurred in A.D. 1346, when Tatars besieging Italian traders in a citadel in Crimea threw corpses of plague victims over the wall, starting a vogue in siege tactics that lasted as long as the Black Death itself did.

Developments in biological warfare and bioterrorism have inevitably kept step with developments in medical science generally. In this respect Pasteur, Koch, and Ehrlich opened Pandora's box when they made the discoveries that gave birth to modern microbiology. With the promise of miracle germ cures came the threat of horrific new germ killers.

In World War I chemical warfare with poison gases was the other side of the coin to the magic bullets of Dr. Ehrlich's chemotherapy. The casualties of these weapons totaled 800,000 men on both sides. In World War II all the major powers conducted germ-warfare research, and some countries went further. Josef Mengele injected Jews and other prisoners at Auschwitz with typhoid and tuberculosis germs. The Japanese formed a large unit for biowarfare in 1936. Based in Japanese-occupied Manchuria, in northern China, the noble-sounding Epidemic Prevention and Water Supply Unit, or Unit 731, experimented on captured Chinese, Soviet, and American soldiers. During the late 1930s and early 1940s the Japanese dropped "disease" bombs on Chinese cities with an unknown loss of life.

The United States did not deploy such weapons, but it had them. It manufactured anthrax and botulinum bombs in the expectation that they might be needed to retaliate against Germany's or Japan's use of biological agents. At the end of the war, conscious that biological weapons could well become more important, the United States shielded Japan's Unit 731 from prosecution for war crimes in exchange for its research data and expertise, just as Hitler's V-2 rocket scientists, such as Wernher von Braun, were brought into America's missile program.

After World War II, a biological-arms race proceeded in parallel with the nuclear-arms race, except that biological weapons capability spread around the world farther and faster because of its relatively low cost and technical demands. To take two recent examples of this ongoing trend, in 1988 Libya built a chemical-weapons plant in the guise of a pharmaceutical factory. And Iraq used mustard gas in its long war with Iran, and has used both mustard gas and toxic nerve agents against its own dissident Kurdish population. Saddam Hussein's stockpile of weapons of mass destruction is thought to include many other germ-warfare agents, as well.

Throughout recent decades, efforts have been made to limit or ban the use of biological weapons. The 1972 Convention on the Prohibition of the Development, Production, and Stockpiling of Biologic and Toxic Weapons and Their Destruction has been signed by one hundred countries, including the United States. But if Saddam Hussein's example is any indication, it has been honored at least as much in the breach as in the observance. Often this occurs with the cooperation of companies in the West that sell so-called dual-use technology to countries like Iraq and Libya. A plant for making pesticides can readily be turned into one for making biological weapons, for example. In the United States, a 1996 law, the Antiterrorism and Effective Death Penalty Act, increases penalties for development of biological weapons or misuse of germs to spread disease. Under this law, authorized germ-culture repositories, like the American Type and Culture Collection in Washington, D.C. , must follow strict protocols to verify that the use of germs is legitimate before shipping any cultures. Likewise, some scientists have suggested that it be made illegal under international law to participate in any biological or chemical weapons activity that violates international treaties.

The pace of biological-weapons development may well be accelerating, thanks to innovations in genetic and molecular engineering. In 1997, *The Cobra Event,* a novel by Richard Preston, revolved around the release of a genetically engineered mixture of smallpox and common cold virus. Reading it inspired President Bill Clinton to push for increased funding to study and defend against bioterrorism. In an eerie example of life imitating art, a team of Australian scientists announced in January 2001 that they had unintentionally created a killer mousepox virus. Working with a weakened virus that usually makes mice only slightly ill, the researchers wanted to create a mouse contraceptive that would trigger an immune response to the mouse's own eggs. They genetically altered one part of the mousepox virus by inserting a gene that directs the production of interleukin-4, an important immune-system chemical in human beings as well as mice. Instead of acting as a contraceptive, however, the result wipes out a mouse's entire immune system. The mousepox virus cannot hurt people, but the implication of this research is that any virus that does afflict human beings, like one of the common cold viruses, could be turned into a killer virus in the same way.

Or perhaps the smallpox virus will be made even deadlier than in its natural form. Smallpox, which is estimated to have killed half a billion people throughout history, more than all wars and other epidemics combined, has been eradicated in nature. This achievement stands as one of the triumphs of modern medicine. The last recorded case occurred in Africa in 1977, and most countries stopped vaccinating against the disease in 1980. As the capstone of medicine's victory over smallpox, practically all stocks of smallpox viruses and vaccines were supposed to have been destroyed by April 1999. Shortly before that date President Bill Clinton ordered that a collection of smallpox viruses and vaccines should be maintained for security purposes until June 2002. The reason for this was the finding by United States intelligence agencies that other collections were being maintained by Russia, Iraq, Iran, and North Korea, among other countries. If another country or a terrorist group unleashed smallpox on the United States, perhaps in a genetically altered form, we would desperately need a collection of virus strains and vaccines that we could work with to find ways to limit the damage. Speaking to this issue, Nobel laureate Dr. Joshua Lederberg has said, "We have no idea what may have been retained, maliciously or inadvertently, in the laboratories of a hundred countries from the time that smallpox was a common disease. These would be the likely sources of supply for possible bioterrorists."

Evidence to support Dr. Lederberg's fear came to the attention of the world media in June 2000. In Vladivostok, Russia, that month, groups of six- to twelve-year-old children found and played with ampoules of weakened smallpox that had been negligently discarded by an old vaccine collection. Because the smallpox was in a weakened form from which vaccines could have been made, the children survived infection, albeit at the cost of permanently scarred faces. It is interesting to note that the doctors who treated the children did not at first know what to do. Since smallpox had been eradicated in nature more than twenty years before, none of them could recognize the telltale signs of humanity's age-old enemy.

The relative cheapness and ready availability of biological weapons, the potential for genetically engineering designer plagues, all these factors have led the FBI and other law-enforcement agencies to suspect that it is a question of when, not if, bioterrorism takes lives in the United States. Threats to spread anthrax have become fairly frequent recently. Although so far they have turned out to be hoaxes, sooner or later we seem vulnerable to being hit with the real McCoy. In 1998, for example, the FBI arrested two would-be terrorists in Las Vegas, Nevada. The two men said they planned to release anthrax bacilli into the atmosphere, and they did indeed have anthrax, but it proved to be a nonpathogenic vaccine strain. In January 2001, twenty-five or more similar threats were received across Canada and the United States. A typical case occurred at a Wal-Mart in Victoria, British Columbia, where a letter was received that claimed to contain anthrax. Fortunately it did not, but before that could be conclusively established the clerk who opened the letter was given a precautionary dose of ciprofloxacin.

Agroterrorism

Instead of trying to spread disease directly in a population, terrorists might attack indirectly by poisoning a country's food supply.

Agroterrorism, as it is coming to be called, is emerging as an increasing threat. USDA administrator Floyd Horn has said that "a biological attack [on America's crops or livestock] is quite plausible."

Many diseases that take a devastating toll on food animals, such as foot and mouth disease (FMD), do not affect human beings at all. Spreading such diseases would be technically easy and a low-risk strategy for terrorists, who would need to take no special precautions in handling and deploying disease-causing germs. But there would be enormous costs to the victim country. The 2001 outbreak of FMD in Great Britain will wind up costing several billion dollars at minimum, in addition to being profoundly demoralizing for the entire country. The epidemic started on a single farm. It spread with devastating quickness, because the virus that causes FMD can easily be transmitted in a number of ways, including by airborne particles and even by people's shoes tracking dirt from infected areas to other places. Although this outbreak happened naturally and is not linked to terrorism, it suggests the extent of the damage that terrorists could inflict on an entire country, or a large region of a country, simply by infecting the herds on one or a few farms.

Imagine then that a terrorist group or criminal organization wreaked havoc on the agricultural industry in a country or a region of a country. To bring attention to a cause or to extract money, the perpetrators could threaten to do the same thing again in a sort of biological protection racket. This seemed like an outlandish plot when Ian Fleming used it in the mid-1960s in his James Bond novel *On Her Majesty's Secret Service*. Today it seems more and more a genuine possibility. Alternatively, organized criminals might use agroterror to manipulate commodities markets. Advance knowledge that certain crops or herds are going to be tainted could be a means to insure huge profits from trading in futures contracts for other crops and herds.

In this connection it is important to note that the spread of FMD in Great Britain in 2001 was helped along by government cuts in inspection and food-safety measures. Just as the United States became complacent about tuberculosis, as we saw in Chapter 8, and slashed funds for surveillance and treatment only to have multi-drug-resistant tuberculosis (MRTB) threaten to assume

epidemic proportions, so Great Britain slashed budgets for surveillance of its food and livestock. In 1991 Great Britain had forty-
three regional animal health offices and 330 government veterinar-
ians. Today it has half as many offices and many fewer vets, as
well as fewer government labs for disease testing, even though the
threat of food-borne illness and livestock disease has been increasing year by year. It is sobering to realize that the virus that causes
FMD was the first animal virus to be discovered by science, in
1898. A full century later, we still can do little more to halt an epidemic of this terrible disease than to quarantine infected animals
and destroy them, often in the hundreds of thousands.

To avoid such disasters, whether from natural or unnatural
causes, the United States and other countries must remain aggressively on guard against FMD and other easily transmittable animal
and crop diseases. In the United States, existing safeguards may
not be enough to protect against the threats that, in a globally connected world, could arrive at any time at an airport near you. In
April 2001, the U.S. Department of Agriculture and the Federal
Emergency Management Agency conducted an agroterrorism war
game, using computer-generated models, that also involved officials from the Departments of Defense, Commerce, Interior,
Energy, and Health and Human Services. The object of the game
was to figure out how to respond to an outbreak of FMD in a farm
state like Iowa. According to the *New York Times*'s account of the
exercise, it showed that the outbreak would quickly spread to
other states and that stopping it would require "the combined
strength of all federal disaster agencies, including the military."

In light of these threats and evidence of biological-weapons
development in so many countries around the world, including a
number that are bitterly opposed to the United States, our society
urgently needs improved abilities to detect, prevent, and limit
bioterrorism and to treat its victims. In recognition of this, between
1998 and 2001 Congress more than doubled funding for efforts to
combat biological, chemical, nuclear, and radiological terrorism,
from $645 million to $1.5 billion per year. The country's best Protective Response Strategy will surely evolve dramatically as a result
of these expenditures. Among other elements, an effective national
Protective Response Strategy might include the following:

- A coordinated plan is needed that networks all federal, state, and municipal antibioterrorism programs. All "first responders" must be adequately educated and equipped to deal with any biowarfare or bioterrorism event.

- Military and law-enforcement personnel and medical and health practitioners, especially "first responders," should be vaccinated whenever possible so that they can more effectively carry out their respective missions.

- Research and development of vaccines and antibiotics needed to prevent and treat disease caused by biological weapons should be stepped up.

- Vaccination programs for the general public should be instituted to protect against the likeliest biological weapons, such as anthrax.

- The public should be kept informed about germ warfare and should be instructed to report unusual activities to local authorities. "Bioterrorism watches" could be introduced on the model of neighborhood crime watches.

- All physicians and health-care providers should familiarize themselves with the symptoms and treatment of the diseases caused by the seventeen likeliest potential biological-warfare agents.

- A nationwide epidemiological surveillance program should link all medical facilities with the CDC or another assigned federal agency in order to identify clusters of cases that might have occurred. Small clusters may signal a terrorist practicing before a larger-scale act is carried out.

The keynote to all this is vigilance. Any complacency or over-confidence will surely prove fatal, sooner or later. The bottom line is that we must remain on the alert for animal and human epidemics and keep searching for better ways to respond to them, whether they are of natural or unnatural origin. Fortunately the

United States has as yet experienced no deaths from bioterrorism. To date the only documented case of bioterrorism in the United States has been the 1984 incident in which members of the Rajneeshi cult poisoned salad bars in Dallas, Oregon. Although no one was killed, eight hundred people were made sick and fatalities could easily have occurred among the elderly and people with weakened immune systems. The 1993 bombing of the World Trade Center, the 1995 bombing of the Murrah Federal Building in Oklahoma City, and the subsequent horrendous destruction of the World Trade Center buildings in September, 2001 conclusively showed that the United States is not immune to foreign or home-grown terrorists. The next bomb could easily be biological, per-haps one that has been genetically engineered to make it especially lethal.

When West Nile virus unexpectedly surfaced in New York City, public-health authorities periodically held press conferences about progress in fighting the strange and dangerous new virus. At one such conference at NYU Medical Center, United States Sena-tor Charles Schumer termed the outbreak of WNV "the closest example of a dry run for a bioterrorist attack that we could possi-bly face." I agree. And although New York City is probably one of the best-prepared cities in the world for such an attack, I would have to say that it did not pass this test with full marks. The WNV epidemic showed that there are significant gaps in our medical surveillance and response capabilities. To make sure that our pre-paredness keeps pace with the threats to our security, peace of mind, and physical health, New York and the rest of the country must take Senator Schumer's observation to heart and act accord-ingly.

13

Germs for Life

In the century and a half since Louis Pasteur proved that germs cause disease, humanity has been embarked on a great quest to learn the secrets of the germ world. Every generation has pursued that quest in the urgent hope and conviction that its discoveries could signal a final victory over germs and disease. But we must stop vying for a triumph over germs, rather than through and with them. If there is one thing above all to learn from our recent experiences and discoveries—from the tragic flowering of the AIDS epidemic, to the puzzles of antibiotic resistance, to the recognition that germs play a role in many cases of heart disease and cancer and may also be involved in such diseases as Alzheimer's and ALS—it is humility in the face of germs' unrivaled adaptability and genetic resourcefulness. To further medicine's long struggle with disease, we will need a vastly expanded knowledge of how germs function in disparate environments, especially the complex environment of the human body, and we must become better and better at recruiting useful germs to our aid. Beyond that, we must always be prepared for more germ surprises.

We cannot triumph over germs, because the health of the planet Earth and every living creature on it depends on them. As we've seen, germs are truly seeds of life as well as disease. All other living creatures have evolved from germs, and none could survive for long without germs to perform essential functions such as breaking down complex organic compounds into reusable simple elements. With this dependence comes vulnerability. The more we recognize the complexity of our interactions with germs, the more we see how easy it is for the balances between germs and people to be disturbed.

This fact has been brought home to me again and again in my career as a microbe hunter. A decade ago, for example, I found the fingerprints of a "new" germ killer. I caught the case from a very agitated fellow scientist, Dr. Larry Senterfiet, then the Director of Clinical Microbiology at the New York Hospital–Cornell Medical Center, in New York City. At the time my laboratory was a regional toxin analysis center, so Larry asked that I test a bacterial sample for toxic shock syndrome toxin (TSST-1). The sample was taken from a dying patient with all the classic symptoms that medicine first recognized in women with tampon-related infections in the late 1970s and early 1980s—low blood pressure, fever, skin rash, and delirium; progressing to respiratory distress and liver and kidney failure.

But this patient was a man, the celebrated puppeteer Jim Henson. And the bacterium Larry found was not *Staphylococcus aureus,* which was known to be capable of producing TSST-1, but one from an entirely different family, *Streptococcus pyogenes,* the familiar bug that causes ordinary painful but rarely serious strep throat.

How had it turned deadly? Larry hypothesized that since it was producing all the same symptoms, perhaps *Streptococcus pyogenes* had somehow started to manufacture TSST-1. It was an excellent guess, but Larry's idea was not borne out by my testing. There was no TSST-1 in the sample. Instead I found that the usually mildly noxious *S. pyogenes* was producing special killer toxins of its own.

Invasive strep disease seems to wax and wane periodically, but such a severe "group A" streptococcal infection had not been reported since the turn of the twentieth century. This strain had apparently vanished for some eighty years before cropping up again in Europe and North America in the 1990s. Around the same time, new and more exotic strains of strep seemed to be emerging, notably the so-called flesh-eating bacterium beta-hemolytic streptococcus group A. This strain of group A streptococcus often invades its victims' bodies via trauma such as a wound or even a bruise. No one knows why these strains emerged when they did.

We now call the infection that struck Jim Henson invasive "TSS-like strep" or "streptococcal TSS," because its effects so closely resemble those of toxic shock syndrome. The symptoms come on quickly and, unfortunately, are catastrophic, leading to death in thirty to eighty percent of the cases in a matter of days, with up to fifty percent of the survivors needing to have a limb amputated. Luckily, the disease is quite rare, with only some fifteen thousand cases registered annually by the CDC, and with immediate intervention can be treatable. In addition to analyzing the toxins in the Henson strain, I tested it against a battery of routine antibiotic agents and found it to be pansensitive, meaning that it would have responded to any one of a number of common drugs. Sadly, by the time Jim Henson, a Christian Scientist, received the medical care he needed, the toxins produced by the strep had already done significant and irreversible damage to his organs.

Henson's tragic case exemplifies both the unpredictability of germs and how human choices and behavior, including cultural changes and personal beliefs, can lead to surprising and unwanted germ consequences. The history of the human species offers abundant lessons in this latter fact, but all too often we continue to act as if we are ignorant of it. Indeed, the price of civilization, its primary unintended consequence, is infectious disease. When the first human beings lived as scattered bands of hunter-gatherers, infectious disease was a minor problem. The invention of agriculture changed that forever. As we saw in Chapter 9, slash-and-burn land clearing gave malaria and other mosquito-borne diseases their big break. Likewise, the dense human settlements that agriculture and animal husbandry made possible provided just the circumstances that infectious diseases such as tuberculosis needed to propagate and spread. Many of these illnesses were animal diseases that jumped species as human beings came to live among their original hosts, from mosquitoes to cattle, sheep, horses, and pigs.

This process did not end in the distant past, it is ongoing. In the mid- to late twentieth century, slash-and-burn agriculture and massive land development projects in sub-Saharan Africa, driven by the needs of an exploding population and horribly exacerbated by a series of bloody wars, created ideal circumstances for a simian immune-deficiency virus, to which chimpanzees and other nonhuman primates had long been acclimated, to jump species and infect human beings with unremitting fury. The environmental disruption that brought remote populations of chimpanzees into contact with human beings to create HIV, the cause of AIDS, was seconded by globalizing trends that have linked the newly teeming cities of Africa with the rest of the world on a daily basis. In parallel with this, the same factors have given renewed impetus to malaria and fueled the emergence of antibiotic-resistant strains of tuberculosis and other diseases around the world, as well as released such new germ threats into the human population as Ebola, hantavirus, arenavirus, and West Nile virus.

There is no way to predict the future, but it is surely safe to say that infectious disease will remain one of humanity's greatest problems for a long time to come. For most of human history infectious disease has been the number-one cause of death. The twentieth century's advancements in understanding and coping with germs, particularly in terms of the development of vaccines and antibiotics, have relegated infectious disease to third place in the United States and other developed countries. In the United States, infectious disease is responsible for about twenty-one percent of all deaths, following heart disease, which is responsible for about thirty percent, and cancer, which is responsible for about twenty-three percent. But if the research that is implicating germs in a substantial proportion of heart disease and cancer cases continues to be confirmed by new data, we may well have to reevaluate our figures and once again recognize infectious disease as the worst of killers.

Meanwhile, infectious disease has never lost its number-one status in the world as a whole. One of the consequences is that some ten million children die of infectious diseases every year. The top germ-related killers worldwide are: (1) acute lower respiratory infection, (2) AIDS, (3) diarrheal diseases, and (4) tuberculosis. According to the International Federation of Red Cross, three diseases alone, AIDS, tuberculosis, and malaria, have killed an estimated 150 million people since 1945, compared with twenty-three million people killed in all the wars since then.

The threats from infectious disease have become especially daunting in recent years because of antibiotic resistance, but we need not lose hope in the face of them. Modern medicine has saved millions of lives. It can save many millions more in the future, if we continue to plumb the secrets of the germ world, guard against overconfidence and complacency, and recognize that our health is deeply bound up with the health of the world's poor and disadvantaged. Above all, it is essential that medicine abandon the treatment paradigm of recent decades, which relies far too heavily and reflexively on antibiotics, and replace it with a new paradigm, which includes antibiotics as one of several potentially appropriate therapies. The range of therapies that can provide alternatives to antibiotics is expanding rapidly as a result of new discoveries. To close this book's observations and lessons from a lifetime of microbe hunting, let me give you a survey of the germ-related research that holds the best promise of more effective cures and treatments. Let me begin, however, by recalling once again that an ounce of prevention is worth a pound of cure.

An Ounce of Prevention

If we prevent infections from occurring, we do not have to worry about antibiotic resistance or other problematic aspects of treating them. In terms of both loss of life and money spent, the costliest infectious diseases in the United States are: (1) AIDS, (2) food-borne illnesses, and (3) nosocomial infections, those which doctors, nurses, and other staff give patients in hospitals and other medical settings. Half or more of the deadly or potentially deadly diseases in these three categories could be eliminated by safe sex, proper food hygiene, and hand washing, respectively. Of course, the use of condoms and other safe-sex practices could also eliminate a large proportion of other sexually transmitted diseases, including chlamydia and HPV, which as we saw earlier have been implicated in cervical and other cancers.

Proper food hygiene includes not only the ordinary household kitchen but also the full range of modern food industries, from growing crops and raising livestock to processing and packaging. Although these practices are heavily mechanized and up-to-the-minute in some ways, in others they have not kept pace with the needs of an exploding population in an ever more connected world. Livestock slaughtering, meat inspections, and packaging techniques often leave much to be desired from the point of view of hygiene. In the light of mad cow disease and the 2001 epidemic of foot and mouth disease that began in Great Britain and spread to Ireland and other countries, we should take a rigorous look at current regulations and safeguards to see if they are appropriate. And we cannot afford to forget that air passengers and cargo—whether human, animal, vegetable, or insect—can carry animal, human, and plant diseases from one country to another in a matter of hours.

As far as hand washing is concerned, the hundred thousand deaths per year that, directly and indirectly, can be attributed to nosocomial infections are a national scandal. It bears repeating here that eighty percent of all infections, from common colds and flu to Ebola, are transmitted by touch. The bottom line is that clean hands can be the most powerful weapons for health on Earth.

Dramatic reductions of AIDS and other STDs, food-borne illness, and touch-transmitted diseases are all within our reach. We only lack the will to spend the time and money to grasp them. The necessary first step is public education. I hope this book can play a part in that. But just consider what a national hand-washing education campaign of the same magnitude as the anti-smoking campaign could accomplish. At a minimum it would save tens of thousands of lives and billions of dollars in unnecessary health costs every year.

Other countries need such campaigns, as well. In the Third World there is a special need to educate caregivers and the general populace alike about the dangers of dirty injection equipment. In these countries people strongly associate the wonders of modern medicine with injections, and they prefer to take vitamins, antibiotics, and other medicines in this form. An estimated twelve billion injections are administered around the world every year; ninety-five percent of them are for purposes other than vaccinations, and almost all of these are medically unnecessary. In the Third World a lack of funds for disposable injection equipment and poor medical and general hygiene create circumstances in which dirty needles are commonly being reused even in hospitals and clinics. Such risky injections cause from eight to sixteen million cases of hepatitis B infection every year, along with from two and a half to four million cases of hepatitis C infection and 80,000 to 160,000 cases of HIV infection. It has recently been estimated, in China alone, that two thirds of its 1.26 billion people have already been infected with hepatitis B, compared to about one in twenty Americans. It is estimated that people lose twenty-six million years of life every year to these preventable infections, at a direct financial cost of $535 million a year. (The figure is calculated by subtracting the age at which people die from the average life expectancy for that region of the world.) This is, of course, not just a problem for the Third World; in today's global village, the indirect costs are felt everywhere, sooner or later. Some international health authorities have thus banded together to form the Safe Injection Global Network (SIGN), which strives to provide clean injection equipment, assist with the management of disposable medical wastes, including syringes and needles, and encourage changes in attitude and behavior that will reduce unnecessary injections.

Educating the people of both developed and undeveloped countries about ways to prevent infection can do more to enhance health, at a lower cost, than any form of treatment. But exciting new therapies are also on the way.

Probiotics

As we enter a new millennium, some of the most futuristic, cutting-edge weapons in medicine's arsenal come from nature's first living things, germs. In 1877 Louis Pasteur and his colleague Jules Joubert noted what we would now call an antibiotic interaction between two bacteria, reporting that urine inoculated with "common bacteria," that is, members of the normal human flora, would kill the anthrax bacillus. A few decades later, Jay Schiotz observed that a patient with a staphylococcal throat infection, despite intensive exposure, did not contract highly contagious diphtheria. To test his discovery, Schiotz sprayed suspensions of staphylococci into some diphtheria patients' throats and reported that this treatment apparently cured the disease.

Since then scientists have come to understand some of the mechanisms by which one species of bacteria might counteract another. For example, *Staphylococcus* germs and *Streptococcus pneumoniae* (a common agent of pneumonia) oppose each other because the pneumococci produce peroxide that inhibits the staph's growth. In 1972, Henry Shinefield made the fascinating discovery that the ability of one strain of *Staphylococcus aureus* to interfere with the growth of a different strain depended on where they encountered each other on the body.

These discoveries, among others, laid the groundwork for a whole new range of "probiotic" therapies (employing live germs to suppress other germs) as opposed to "antibiotic" (germ-killing) therapies for many infectious diseases. While medical traditions in Europe and Asia have long espoused the benefits of probiotic therapy, the United States is just now coming to recognize the tremendous potential it holds. First of all, live "good" germs can control many "bad" disease-causing microbes that have grown resistant to antibiotics. Second, they are proving effective against diseases that antibiotics cannot effectively treat. To cite just one example, according to some studies beneficial lactobacillus germs in vaginal suppositories may check the stubborn urinary-tract infections that plague many women; in yogurt or in beverage form, they can help combat the diarrhea-causing rotavirus infections that afflict billions of children worldwide. Preliminary evidence suggests that other probiotics may help prevent the *Helicobacter pylori* infection that causes ulcers, lower blood lipids (by converting cholesterol to coprostanol, thereby reducing intestinal absorption) and high blood pressure to protect against heart disease and strokes, and possibly even reduce or inhibit the development of precancerous colon lesions by either detoxifying carcinogens or suppressing growth of intestinal bacteria that possess carcinogen-generating enzymic activities.

Here is a detailed example of how probiotics can work. Most of us have experienced diarrhea during antibiotic therapy. Although the bout of diarrhea is usually short-lived, it can become more serious if therapy is prolonged or the diarrhea is caused by a highly antibiotic-resistant bacterium called *Clostridium difficile.* Usually *C. difficile*-associated diarrhea is related to an antibiotic's upsetting the normal microbial ecology of the gastrointestinal tract. The antibiotic kills off the normal good flora and selectively favors the potentially pathogenic *C. difficile,* which produces a powerful toxin that causes diarrhea. This occurs quite often in the hospital setting, where many patients are treated with heavy doses of antibiotics. The clinical spectrum of C. *difficile* infection can range from an asymptomatic carrier state, to simple diarrhea, to life-threatening colitis. Up to forty-six percent of adult hospitalized patients may harbor C. *difficile* in their intestines. So it is no wonder that up to twenty-one percent of adult patients, and up to thirty percent of all patients, become infected with C. *difficile* if they are hospitalized for more than two days. It is estimated that this illness adds as much as $5,000 to the cost of a hospital stay.

To control C. *difficile* infections, different antibiotics can be tried. But that approach is costly and inefficient, and it ultimately furthers increasing antibiotic resistance. Lactobacilli offer a better way. A strain called *Lactobacillus casei* GG can survive the acidity of the stomach when it is swallowed in tablet form. It is a highly concentrated relative of the lactobacillus bacteria that are found in yogurt, and even better able to compete for space in the intestines with C. *difficile* and prevent that bacterium from taking over. The lactobacilli hold the fort, until the antibiotics are stopped and the normal intestinal bacteria can reestablish themselves.

The time has come for more careful exploration of the therapeutic applications of probiotics against a larger spectrum of infectious diseases. Probiotics should play an increasingly important role in the treatment of infectious diseases in the decades ahead.

Antisense Drugs

Just about all human disease is the result of destructive or inappropriate protein production. This is true both for infectious diseases such as toxic shock syndrome and noninfectious diseases such as cancer. For example, many pathogens, like the flesh-eating bacterium beta-hemolytic strep, can produce protein enzymes that help it invade and destroy human body tissues, which in turn enables the bacterium to produce protein toxins which poison the body, giving rise to severe infection or even death. Antibiotics can interfere with germs' protein production, but antibiotic resistance creates the need for other treatments. Antisense drugs are a promising alternative. They are designed specifically to inhibit production of particular disease-causing proteins. Antisense drugs work in the genes, whose DNA contains the necessary information to produce proteins.

The DNA molecule was famously dubbed "the double helix" by its discoverers Francis Crick and James Watson. In other words, it is made up of two complementary strands of DNA entwined in the shape of a corkscrew. When a protein is made, the two complementary strands of DNA partly uncoil. One strand is called "sense" and the other is called "antisense." The antisense strand is used as the template for special enzymes, which assemble messenger RNA by a process called transcription, and the RNA in turn makes a specific protein from amino acids within the cell by a process called translation. The antisense drugs are designed to complement the messenger RNA, binding to specific regions on it in order to prevent protein production. Antisense drugs can be made to prevent protein production of any cells, including human cells. The effectiveness of the antisense technique has been proven in the laboratory and in animals and human beings, and new antisense drugs are being developed against a wide variety of infectious, inflammatory, and immune diseases, as well as cancer. Antisense drug therapy alone could revolutionize the treatment of disease.

Decoys and Blockers

In order for germs to cause disease, they must first attach to the cells of a host organism. In the case of common flu viruses, for example, viral particles have receptor sites that enable them to attach to areas of sialic acid on cell surfaces. Given this fact, it would obviously be desirable to develop some sort of decoy receptor site for the flu virus in order to trap it before it can infect a host's cells. Just such a decoy therapy was recently developed at the University of Michigan using an artificially created decoy made from a nontoxic polymer to which sialic-acid receptors were attached. This decoy system is about the same size as ordinary host cells, so that it can compete effectively with the sialic-acid receptor sites on body cells. The virus attaches to the sialic acid on the polymer, gets locked into place, and thus cannot replicate.

Decoys should slow down or prevent infection if they are introduced into the human body. Initial laboratory experiments with decoys have shown that they can work, and new experiments using mice are under way. Decoy therapy should prove useful against a wide variety of infectious germs, including the AIDS virus. Decoys could be distributed throughout the entire body or clustered at body sites where germs typically try to gain entry, such as the nose and throat for cold and flu viruses.

Some viruses may escape the decoys, and then manage to infect host cells and replicate inside them. But the decoys will be ready and waiting for the next generation of viral particles to be released into the bloodstream by the lysing, or bursting, of the infected host cells. Eventually most if not all of the viral particles will become inactivated by the decoys, giving the immune system a big assist.

Similar thinking underlies efforts to create blocking agents that can be introduced to prevent germs from latching on to host cells. For example, to infect human cells, HIV particles must latch on to one receptor and one coreceptor site in each of two immune-system cell types, namely macrophages, which HIV uses to transport itself throughout the body, and T cells, which HIV destroys. A small group of mostly caucasian people, about one percent of the population, lack one of these coreceptors and so seem to be immune to HIV. The search is now on for molecules, dubbed entry inhibitors, that can block one receptor site and stop HIV infection in its tracks.

Multiheaded Antibiotics

The first question with regard to antibiotics use will always be to make sure that they are really necessary. Cutting down on unnecessary use of antibiotics is the single most important thing that can be done to fight antibiotic resistance in germs. Although that self-policing will remain an urgent necessity, medical science is exploring ways to make antibiotics more effective. As we saw in Chapter 11, there are five general mechanisms that antibiotics use to attack germs. Germs can develop and share resistance traits to each of them, but the odds will be strongly against any one germ's acquiring resistance to all five. These probabilities present an opportunity for a multiheaded antibiotic, one that, like a multiple armed missile, can fight germs in more than one way.

In bacterial cells mutations occur, on average, about one in every million cell divisions. Mutations can occur more frequently, but we can take this rate as a reasonable baseline and assume that a mutation conferring resistance against a hypothetical antibiotic is likely to arise in a hypothetical germ sometime in the course of a million cell divisions. Given how quickly most germs reproduce, that one-in-a-million chance can easily occur in the course of antibiotic therapy. But if an antibiotic were able to inhibit bacteria using two of the five possible attack mechanisms, then the chances for developing a mutation that would confer resistance would be one in a trillion (1×10 to the twelfth power). Similarly, if an antibiotic were able to inhibit bacteria using three of the five mechanisms, then that would mean resistance would have a probability of occurring after one thousand trillion (1×10 to the eighteenth power) cell divisions. In other words, threefold antibiotic resistance would take a very long time to develop, and would be highly unlikely to occur before a three-headed antibiotic could do its work. If an antibiotic were able to inhibit or kill bacteria using all five mechanisms, then resistance would have a probability of occurring after 1×10 to the thirty-second power cell divisions (that is 1 with 32 zeros after it). Of course, there are other means by which bacteria can develop resistance to antibiotics, such as by acquiring plasmids. Nevertheless, it remains improbable for a bacterium to develop resistance to all of the mechanisms offered by a multiheaded antibiotic during therapy.

At present an antibiotic usually has only one bacterial target site that it acts on. Because antibiotics typically have only one specific site of action, they are readily overcome by resistant mutations. Several pharmaceutical companies are developing multiheaded antibiotics that will challenge even the most resourceful antibiotic-resistant germs. Thus, if one of the antibiotic's warheads, so to speak, is defused by a resistance gene inside a germ, it will have other warheads to deploy to kill or inhibit the germ. In addition to being multiheaded, the new antibiotics under development will work in novel ways. They will be aimed at preventing adherence and colonization of germs to human cells or their expression of other virulence factors such as, among others, pili, capsule, enzyme, or outer surface protein production.

Genetic Engineering

Genetic engineering, which has made it possible to alter specific traits in the genetic makeup of an organism or a cell, began with germs. In 1971, as an outgrowth of research done in Israel, scientists inserted genes into certain bacteria that gave them the ability to break down the crude oil found in tankers like the *Exxon Valdez*. Since then scientists have eagerly applied the same techniques in other areas, and today one-quarter of all plant crops grown in the United States—including half of the soybeans and cotton and one-third of the corn—are genetically engineered, a fact which makes the current debate about genetically engineered food at least in part a case of shutting the barn door after the cows have gotten out.

Like all other living things, all germs contain chemicals called DNA (deoxyribonucleic acid) and RNA (ribonucleic acid). (Viruses are only an apparent exception. Viruses contain either DNA or RNA, but not both, which makes them unable to reproduce on their own and forces these particles to invade cells and hijack their reproductive machinery in order to replicate; for this reason viruses are, technically speaking, not alive.) Within the DNA in every living thing is contained all of the information that an organism or a cell needs to maintain its life and reproduce itself. Placed end to end, the coiled DNA in a human body would be long enough to reach the moon. No two creatures have identical DNA except for identical twins and clones.

All of an organism's DNA resides in distinct organized structures called chromosomes. Each chromosome is made up of genes. The specific sequence of DNA in each gene carries information for constructing proteins including enzymes, which are essential for life. I said that all human disease is the result of destructive or inappropriate protein production. This is because all living organisms are composed largely of proteins. You might say that at the most basic level, all living things are just protein synthesizers. Any activity that involves cells, whether in a creature's brain, heart, or gut, depends on protein synthesis that is coded for in that creature's genes. The sum total of all the protein synthesis that an organism is capable of is its genome.

Human beings can synthesize over eighty thousand different proteins. To carry the codes for this activity, the human genome turns out to require approximately thirty thousand genes on twenty-three distinct chromosome pairs. This is an enormous number, in terms of the gene interactions it makes possible, but it is nothing like the hundred thousand genes that many scientists confidently predicted in comparison to the twenty thousand of the lowly roundworm. This tells us that numbers alone will not account for the enormous obvious differences between human beings and other living things. But it in no way reduces the complexity of explaining what does account for them. Human beings' complexity arises not from the sheer number of genes, but from the dynamic interaction of those genes and the proteins they synthesize in an intricate structural network.

People who thought that sheer numbers would be a more significant part of the answer than they apparently are were disappointed to learn that human beings have only about thirty thousand genes. In this connection it is important to recognize that the new science of genetics is in its infancy. Although it can already boast many breakthrough achievements, some scientists and others want genetic science to run before it can walk. The excitement over cloning exemplifies this. Talk of cloning human beings is much in the air, now that sheep, mice, pigs, and various other animals have been cloned, but science has far to go before it can replicate a human being's unique genome with any reasonable hope of success. Fewer than three percent of cloning experiments succeed, because a cloned genome is unstable and much more prone to breaking down than the original. At present we seem better able to clone freaks and monsters than supermen.

The debate over cloning is matched by that over genetically engineered food. On one side of the debate are those who hope for higher-yielding, more disease-resistant plants and healthier, more nutritious food. On the other are those who fear poisonous "Frankenfoods" and lab-created marauding plants that will elbow out natural plant stocks and change an ecosystem for the worse.

I believe we can find our way to realizing the positive opportunities represented by genetic engineering, but it will probably take many slow, careful steps to get there. Along the way we will have to guard vigilantly against overconfidence that could lead to terrible harm. At the same time it has to be said that genetic engineering is enjoying a wonderfully productive infancy. We are quickly learning the lay of the genetic landscape and figuring out where to look for answers to inherited disorders such as muscular dystrophy, cystic fibrosis, diabetes, and sickle-cell anemia, among others, and to predispositions to diseases such as cancer. This is making it possible to identify the genes responsible for specific disease traits and to work toward finding ways to alter them to restore or maintain health.

In this effort germs stand ready to guide and help us. The astonishing array of genetic tricks that germs employ to infect a host organism and thrive inside it constitute an encyclopedia in a secret language that scientists are translating at a feverish pace of discovery and experimentation. By identifying the genes responsible for microbes' actions, scientists can begin to "create" germs that can perform such tasks as eating toxic waste or fighting other, disease-causing germs. Surrogate viruses and bacteria can even carry germs to a designated part of the body, one of several techniques associated with gene therapy.

Genetic Therapy

Surrogate germs can be used to deliver genes that an individual may be deficient in. For example, a French research team reported the first successful gene therapy on three babies that were saved from certain death from a syndrome called severe combined immune deficiency (SCID-X1). Such children can survive only in a germ-free bubble; outside they face certain death from infection by the myriad germs that are present everywhere in the environment. The French researchers extracted some bone marrow from each of the three babies. From the bone marrow they isolated particular cells called stem cells, which contained the defective gene that causes SCID-X1. These cells were mixed with a harmless virus containing a healthy form of the gene. The virus then infected the stem cells, acting as a vector for the healthy normal gene. The cells containing the healthy gene were then transplanted back into the babies. These genetically corrected cells grew and eventually they replaced the cells with the defective gene. The result was that the "bubble babies" acquired immune systems that function normally. The success of this lifesaving experiment opens the door to a wide variety of possible cures of both immunologic and nonimmunologic diseases using the same gene-therapy techniques. Diseases such as hemophilia, cystic fibrosis, Alzheimer's, diabetes, and even cancer may one day meet their defeat at the hands of gene therapy.

One very interesting surrogate virus being tested in an effort to develop a method of repairing spinal-cord injuries, such as the one suffered by the actor Christopher Reeve, is the poliovirus. Yes, the poliovirus. The same virus that crippled two million Americans in the past is now being used to help mankind. A research group at the University of Alabama has modified the poliovirus by deleting the genes responsible for causing the disease polio. But these researchers left intact the virus's ability to invade the motor neurons. The logic behind this is that the poliovirus can be given beneficial genes, which can then be delivered directly into the spinal-cord motor neurons, where they can, it is hoped, repair nerve cells. In this case the poliovirus also becomes a vector. This procedure has already been tested successfully in mice, and this is exciting news indeed for patients with spinal-cord injuries, which in the past were thought to be irreparable. Isn't it a remarkable turn of events? The very viruses responsible for destroying nerve cells are being harnessed to repair them. Two other research groups, one from Duke University and the other from Southern General Hospital in Glasgow, have genetically designed a poliovirus and Herpes simplex virus respectively to help destroy gliomas, which are tumorous brain cancers that are nearly impossible to treat. Preliminary experimental results in mice, as well as more advanced human studies, have been encouraging.

Along the same lines, a Hungarian research group has reported success in establishing the first effective methodology for generating an artificial human chromosome from predictable sequences. Such a chromosome could provide a good vector for gene-therapy purposes.

Biopharming

Biopharming is the process of genetically altering plants so that they grow to contain an edible vaccine or some other edible pharmaceutical. Imagine vaccine-, antibiotic-, or medication-rich foods that replaced pill or injection time with mealtime, and it is easy to see why a brand-new industry is emerging to exploit the potential of biopharming. Over two dozen companies worldwide are now working to transform plants into miniature drug-production factories. Among the drugs being produced in this way are industrial chemicals, enzymes, and even vaccines for infectious diseases.

Believe it or not, many of these drugs have already been tested on human beings in clinical trials with promising results. One company is currently testing a potato with a Norwalk virus vaccine, and another experimental potato contains a vaccine for the hepatitis B virus. Eating either, theoretically, will allow an individual to develop antibody protection against disease.

A group of scientists from Thomas Jefferson University in Philadelphia are working on delivering an AIDS vaccine in a bowl of spinach. Because delivery of a plant-based rabies virus vaccine has already been reported to be successful, the Philadelphia group is hopeful that they can produce the same result by using this method with HIV. Providing drugs in plants is far cheaper than factory production. Drugs produced in factories can cost thousands of dollars per gram, while drugs produced in plants can cost as little as manure or fertilizer. This holds especially great promise for developing nations, which are often unable to pay the high costs of medications patented and produced by the world's large pharmaceutical companies. Biopharming could go a long way toward reducing infectious diseases as the number-one killer of people worldwide.

Botanical Drugs

In many cultures herbal medicine remains the primary method used to treat both infectious and noninfectious diseases. Although we do not think of modern Western medicine as depending on herbs, about half of all drugs used in the United States have their origin in plants, including many antibiotics derived from fungal or bacterial sources. In addition, plant extracts, or botanicals as they are called, have become very popular alternative medical treatments for a wide assortment of ailments. Because of antibiotic resistance, pharmaceutical companies and other investigators are once again searching for antibiotics and other medicines in plants and soil microbes, and mainstream medicine has become more receptive to the use of antimicrobials derived from plants. For example, researchers at both Tufts University and Colorado State University discovered that a compound from a common lawn shrub, the barberry bush, can help an antibiotic to overcome antibiotic resistance.

A related finding is that extracts from the fruiting bodies of the Japanese tree *Terminalia chebula* inhibit *Staphylococcus aureus*. What is especially useful is that this extract is effective against antibiotic-resistant as well as antibiotic-susceptible strains of *S. aureus*. An extract of the licorice plant *Glycyrrhiza* has been found to be effective against the HIV virus in mice. In Guatemala, women with vaginitis caused by the yeast *Candida albicans* were treated with extracts of a common native plant called *Solanum nigrescens*. When the women were given vaginal suppositories of the plant extract, they were as effective as the commonly used antifungal drug called nystatin. A recent study of a group of elderly women confirmed that individuals drinking cranberry juice have fewer bacteria in their urine than untreated controls. And various peptide extracts from barley, wheat, and even fava beans have been found to inhibit a wide range of bacteria such as *E. coli* and enterococcus. Similar compounds in the sugar beet are active against various fungi. The possibilities go on and on. All over the world scientists are reexamining plants for their antimicrobial potential. This is an exciting new/old area of research that offers the promise of major discoveries.

Bacteriophage Therapy

In 1915, Felix d'Hérelle, a Canadian microbiologist working at the Pasteur Institute in Paris, discovered tiny bacterial viruses that lysed, or burst, bacterial cells. D'Hérelle called these viruses bacteriophages, which literally means "bacteria eaters." He predicted that his discovery could revolutionize the treatment of infectious diseases. Almost a century later, the recognition that he was right is finally becoming widespread. In a 1917 research paper, d'Hérelle described his isolation of an invisible microbe that was endowed with a unique property. It was able to kill the dysentery bacillus (*Shigella*) that was causing an infectious-disease problem of his day. D'Hérelle correctly concluded that specific bacteriophages would act on specific germs. In fact, there are hundreds of different types of phages, each one of which kills a different germ.

Bacteriophages are viruses that infect bacterial cells. When a virus infects a cell, either bacterial or human, it takes over the cell's reproductive machinery and then usually lyses the cell to release a new crop of thousands of viral particles, which can infect other like cells. Bacteriophages reproduce quickly. One bacteriophage can produce tens of thousands of viral particles in one hour. Hence over time bacteriophages can kill off all of the susceptible bacterial cells they come into contact with.

D'Hérelle discovered that patients who were infected with *Shigella* quickly got better if a bacterial phage was also present. He purified bacteriophages and successfully used them to treat infections. Such therapy theoretically could be used to treat any bacterial infection so long as the right phage were used. Although the American pharmaceutical manufacturer Eli Lilly produced and marketed therapeutic phages in the 1930s, problems associated with their purity delayed their widespread use. With the success of penicillin and other newly discovered antibiotics during World War II, the pursuit of phages fell out of favor, except in the Soviet Union.

The Soviets' continued interest in phages was lucky for the world, because the advent of multiple-antibiotic-resistant microbes, so-called super bugs, has once again made phage therapy attractive. Soviet scientists systematically developed phages after World War II, and Soviet doctors learned to mix and match them with great skill. The Eliava Institute in Tbilisi has the largest phage library in the world, with more than three hundred phage types, which can be used to treat all types of infections. For example, gastroenteritis can be cured with a cocktail of seventeen different phages, burn-wound infections can be treated with other blends, and so on. Even antibiotic-resistant strains of *S. aureus* will succumb to phage therapy. Numerous American entrepreneurs are now interested in getting into the phage business as a way to profit from antibiotic resistance.

The great advantage of phage therapy is that a specific phage will kill only a specific germ. Antibiotics, in contrast, kill many beneficial germs and innocent bystanders in addition to their intended targets. One extraordinary application of this has already gotten under way. Researchers in the U.S. have developed a strategy for incorporating phages against *E. coli* 0157:H7 into cattle feed in hopes of eliminating this pathogen from the food chain. The first step toward this goal was to select a highly specific phage, which would target only the 0157 serotype, leaving the normal flora of cattle's gastrointestinal tracts undisturbed. In a like manner, development of phages specific for *Salmonella* have been developed. These phages will be given to poultry in order to eradicate that pathogen in these animals. Some investigators have pointed out that bacteria may develop resistance to phages; in order to eliminate that possibility, multiple phages will be used. This will more assuredly guarantee complete eradication of the pathogens and virtually eliminate the possibility of resistance development. Indeed, phage therapy is yet another new/old alternative antimicrobial therapy which shows a great deal of promise.

Immune Modulation

That there is an interrelatedness between the endocrine system (hormones) and the nervous system has long been known. Recently there has been intense interest in the interaction of both hormones and the nervous system with the immune response to disease. Many scientists have reported that when the body's immune response springs into action against an intruding germ invader, it causes production of certain inflammatory substances. These chemical substances produced by the immune system are detected by the central nervous system (CNS), which monitors their level. When the CNS detects a particular threshold level, it causes the release of certain hormones, which downregulate the immune response, thereby acting as a control to prevent a hyperimmune response that might itself injure the body. A system of checks and balances, if you will.

Hormones have also been shown to directly modulate the immune response. For example, why don't the immune cells of the female body attack sperm cells entering the uterus? Surely they are foreign to it. Researchers have shown that hormones produced during the menstrual cycle actually favor an infection-free environment in the uterus prior to ovulation, which controls germs yet doesn't ordinarily kill sperm or the implanted fertilized egg. Further, scientists have shown that every type of immune cell has its function tightly regulated by hormones during the menstrual cycle. Current evidence suggests that the main hormonal regulator of this immune function in females is progesterone, and analogously in males is testosterone.

An awareness of the interrelatedness of the endocrine, nervous, and immune systems is helping scientists to unravel the complex factors that govern the state of health and disease, which in turn will help to cure infectious and immunological diseases in the future. It should also point the way to enhancing immune responses to different diseases. In other words, this knowledge will enable us to modulate or fine-tune our immune response. This prospect has opened up a whole new field of study of the three-way communication between the nervous, endocrine, and immune systems. Preliminary studies have already shown great promise for the treatment of autoimmune diseases, cancers, infectious disease such as AIDS, physiologic disorders such as shock, and many other illnesses. We have only begun to explore this holistic modality of treatment.

Nanotherapy

Imagine millions of nanocomputers (nano means "dwarf" or "one-billionth"; a nanometer, one-billionth of a meter, is roughly the width of six carbon atoms) injected into the body. Hundreds of such nanocomputers, dubbed "nanorobots" by Robert A. Freitas, one of the scientists working on their development, could fit inside a single biological cell. Inside the bloodstream, naonorobots could act as sentinels, detecting cellular anomalies or defects, and also as clean-up crews, destroying invading microbes and repairing cells damaged by genetics, germs, or physical or chemical trauma.

The prospect of nanotherapy may remind you of the classic 1966 science fiction movie *Fantastic Voyage,* in which an experimental submarine and its crew—played by Raquel Welch, Stephen Boyd, and Donald Pleasance, among others—are miniaturized in order to perform internal brain surgery on a patient. But the beginnings of nanotherapy are already science fact. For example, an engineer at Cornell University, Carlo Montemagno, has succeeded in creating a nanomotor that simulates the flagellum that many bacteria use to propel themselves in a medium like the bloodstream. By the middle of this new century, nanorobots and nanomedicine may indeed be realities.

Nanotherapy will then take its place in a world of nanotechnologies that will connect people and information as never before. We can see signs of this already in the proliferation of wireless Internet-ready devices that manufacturers are installing in cell phones, household appliances, and other products. In the nano-age, every person on the planet may have 24/7 access to the Internet through a supercomputer that fits in a button or, "calling Dick Tracy," a wristwatch. Eventually the division between the computing power of human brains and that of manufactured devices may blur into a seamless whole, allowing large groups of people to unite mentally, for good or ill, into symbiotic super-organisms. If that happens, humanity will have evolved full circle from its origins in a single germ cell floating in the primordial soup some four and a half billion years ago.

New-Wave Vaccines

The concept of vaccination arose in ancient times. Chinese folklore alleges that smallpox vaccination occurred as early as the sixth century, although the first written record appears to be attributed to a Buddhist nun practicing during the reign of Jen Tsung (A.D. 1032 to 1063). She recommended selecting scabs from victims and letting them dry for about a month, then grinding them with particular plants and blowing the mixture into the nostrils of healthy individuals. Similar methods were employed by sixteenth-century Hindus in India. But notwithstanding these historical anomalies, the British physician Edward Jenner gets the credit for the first scientific attempt to control an infectious disease by means of systematic inoculation (vaccination).

In 1796, Jenner demonstrated that individuals injected with cowpox (a disease, similar to smallpox, that infects cows, but not people) were protected against smallpox. Years later, Jenner admitted that he had been inspired to do this by reading about an English cattle breeder by the name of Benjamin Jisty, who became immune to smallpox after contracting cowpox from his herd, and who then inoculated his wife and two children with cowpox in order to avoid a smallpox epidemic. The technique that Jenner employed in his experiments came to be known as vaccination, from the Latin word *"vacca,"* meaning cow. Louis Pasteur actually coined the word in honor of Jenner after applying his concept of injecting a weakened form of a pathogen into an animal, which thereby gained immunity against the more virulent form of the germ. Pasteur used a weakened form of rabies in his experiments.

Vaccination is common nowadays and has offered protection to most people in the industrialized world against eight childhood diseases: smallpox, DTP (diphtheria, tetanus, and pertussis), poliomyelitis, measles, mumps, and rubella. In addition, there are currently new vaccines against influenza, hepatitis B, pneumococci, and *Hemophilus influenza* type b. Vaccinations without a doubt have had a greater and more significant effect on reducing death in the world than even antibiotic therapy. With the growing antibiotic-resistance threat, vaccination has assumed an even more significant role in the battle against infectious diseases.

Pasteur's rabies vaccine is perhaps the most celebrated in medical history. He used a weakened (attenuated) rabies virus grown in the brain tissue of rabbits in order to prevent hydrophobia (rabies) in human beings bitten by rabid animals. Many objected to the idea of injecting such a deadly virus into people, especially after word got out that some people experienced serious postvaccination reactions, including encephalitis. We now know that these complications were a result of impurities, as well as people's individual immune responses. Today we try to minimize such side effects by purifying vaccines as much as possible, and in spite of occasional complications, vaccines have proven their worth beyond any measure.

Probably the golden age of vaccines began in 1949 with the development of poliovirus vaccines using cell cultures rather than animals, in order to grow the viruses in purer form. Purification techniques have gotten even better over the years. The current focus is on sub-units (purified proteins or polysaccharides) of germs or genetically engineered products rather than live germs, which pose a greater risk of illness and unwanted side effects.

The deciphering of the human genome provides a model for deciphering the genome of every known pathogen, and knowing the DNA of a germ can lead to a better vaccine. For example, a new measles vaccine involves injecting DNA from a vaccine virus's nucleic acid into the body of a child. Unlike traditional immunizations, which use weakened or killed viral germs that can still cause illness in a small number of cases, such a DNA vaccine poses no threat of infection. Many disease-causing germs enter the human body via a mucosal surface in the respiratory, gastrointestinal, or urogenital system. To fight them, mucosal vaccines are now being developed using the same DNA-based strategies or the utilization of cellular sub-units in combination with numerous other elements.

The new-wave vaccines could eventually counter infectious disease agents, such as HIV, that are not now controlled by antibiotics or other drugs or measures. According to the CDC about eighty-five hundred individuals (about seventy-five hundred adults and a thousand children) become infected with HIV every day. As of the year 2001 approximately forty-two million individuals have cumulatively become infected with the HIV virus. The simple truth of the matter is that adhering to a drug-taking regimen that includes up to seventy pills a day will not stop the worldwide AIDS epidemic. Only a vaccine can do that.

Thus it is indeed fortunate that the potential for vaccine development has never looked brighter. If vaccines could be developed in the near future for only three major killers of people, namely AIDS, tuberculosis, and malaria, that would save approximately five million lives each year, especially in poor countries where existing drug therapies are prohibitively expensive. There could be no greater gift to humanity.

Until these tantalizing new therapies come on line, of course, antibiotic- and drug-resistant germs will continue to pose deadly threats. But with Protective Response Strategies grounded in common sense and good science, we can do much to maintain our health in the meantime. An overarching **Protective Response Strategy** against infectious germs would incorporate the following concepts and behaviors.

First, we must understand the important role that germs play as recyclers of complex organic matter on planet Earth, and in the maintenance of our own health. Germs in the normal human flora can prevent establishment of pathogenic germs in our bodies and also provide us with nutritive materials. Therefore, we must never entertain the notion that all germs are bad and should be eliminated. That is neither possible nor desirable. Because germs are ubiquitous (including in and on human beings), we must learn to live with them. Hence our protective strategy should be focused on reducing the risks of infection.

We must also maintain an awareness of the dynamics involved in the transmission of germs. Such knowledge can lead to better methods of control or even prevention of infectious processes. We should recall that the causative agents of disease are usually transmitted from a so-called reservoir of infection to a susceptible host by four main routes: direct or indirect contact (such as hand to hand, or hand to doorknob), common vehicles (such as food- or water-borne germs), airborne (such as inhaling tiny particles in air) and vectors (such as flies, ticks, or mosquitoes). Knowing this may better enable us to defeat or prevent infection and help us establish effective defensive strategies.

The infectious process is in a dynamic state of flux and involves the interrelatedness of the germ, the host, and the environment. One way or another, this means that infectious germs are a 24/7 threat. We should therefore practice infectious-germ avoidance strategies until they become second nature. And because weaker hosts may succumb to germ threats that leave stronger individuals unscathed, it is important to get regular exercise and eat a properly balanced diet.

Thus it is essential to maintain a commonsense attitude regarding prevention of germ-caused diseases. For example, if someone with obvious flu-like symptoms touches a doorknob that you were about to touch, stop. There's no need to risk being infected with the flu, or a deadly illness whose early stages mimic it. Get a tissue or other cover and then touch the knob, or use a different exit. If you must open the same door without protection, make sure you wash your hands prior to eating or drinking or touching yourself on any mucous membrane. Always avoid clear dangers.

No medical advance can replace commonsense personal hygiene. Wash hands properly before eating or drinking anything. Never put articles or your fingers into your mouth, or touch a food, until you have first washed them in an appropriate fashion. When you do have contaminated hands, get into the habit of not touching your eyes, your mouth, or your nose until after washing. Hand washing is an especially important preventative strategy because, as I cannot repeat too often, over eighty percent of infections are transmitted by touching. Hand washing is the single most important thing that you can do to safeguard your health.

Always wash fruits and vegetables prior to consumption. Never consume raw meat, fish, or eggs, or unpasteurized cheese, milk, or other beverages. Never eat or drink anything in a low-hygiene circumstance. Sushi and sashimi lovers, please take another look at Chapter 6.

Think of others, as well as yourself, to reduce the spread of infections among family, friends, and other people. If you're sick, stay home from work or school until you are well. Sick children should never be allowed to go to school or to play with other children until they are well. Try to limit handshaking and face kissing to family and close friends. And always practice safe sex!

Finally, seek medical advice when you are ill. Timely treatment can sometimes mean the difference between life and death.

The Most Powerful Life Force on Earth

After all is said and done, it is up to each and every one of us to safeguard ourselves and our families. Although there is much that only government and public-health authorities can do, no surrogate can shoulder the individual citizen's responsibility to exercise common sense about germs, based on a solid understanding of how they function. I hope this book has helped you to gain that understanding, which for me is inseparable from a sense of wonder at the marvels of the germ world. A thousand disparate germ phenomena might vie to be the capstone illustration of these marvels. But here is one that strikes me as particularly intriguing.

In the last few years a number of scientists have observed that bacteria can actually signal each other to take cooperative action. For example, certain marine bacteria that live in a mutualist relationship with squid or other fish can signal each other to glow to illuminate prey or to blind or fool larger predators that might eat their hosts. Other bacteria, including ones that cause infectious-disease processes in human beings, can signal each other to produce toxins that will speed infection and poison a host organism. The fascinating thing about this is that the glowing or the production of toxins only occurs when the bacteria are present in sufficient numbers to have the desired effect. It turns out that the bacteria emit signal molecules, called autoinducers, into their immediate environment, and that receptors on the bacteria allow them to detect these signals from their fellow germs. But no cooperative action occurs unless the bacteria are densely packed together so that the concentration of signal molecules crosses a threshold level.

This phenomenon has been dubbed quorum sensing, because it prevents the bacteria from voting to take action prematurely, so to speak. It has tremendous implications for medicine, offering the prospect of drugs that can prevent infectious germs from doing harm by interfering with their communication system. Beyond that, it gives us a tantalizing clue to the mystery of how complex, multicellular beings like ourselves evolved from simple one-celled germs. We human beings, possessors of language and other forms of symbolic communication like music and mathematics, think of ourselves as the greatest and most efficient communicators in the known universe. But it seems that lowly bacteria have much to teach us about the cellular basis of all the signals we constantly exchange in our daily lives.

We are now coming full circle, from believing ourselves to be at the mercy of nature, mankind's lot since the beginning of time; through entertaining the hope that we could conquer nature with our scientific ingenuity; to recognizing that nature, the magnificent architect, has already devised the remedies for even our most grievous afflictions, if only we will throw ourselves upon its mercy. The young science of microbiology, a mere century old, is now only beginning to scratch the surface of the mysteries that govern the most powerful life force on Earth, germs.

Index

Aedes aegypti mosquito, 183–85
Aedes albopictus mosquito, 185
Aedes (Oclerotatus) japonicus mosquito,
    174–76, 178, 185
Africa, HIV/AIDS in, 113, 114
agent in the chain of infection, 48
agriculture, 134, 253
    agoterrorism, 246–48
    antibiotic use in, 228, 229, 231
    genetic engineering and, 136
    slash-and-burn, 181, 186, 253
AIDS, 16, 71, 112–14, 149, 151, 154, 157,
    164, 182, 207, 251, 254, 255,
    256, 267, 275
    tuberculosis and, 157–58, 159
    see also HIV
airborne spread of germs, 52–53, 54,
    149–69, 275
air-filtration devices, 169
airline travel:
    airborne spread of germs and, 52–53, 161
    mosquitos transported by, 175–78, 185
Alcott, Louisa May, 119
allergies, 69, 72, 163, 164, 165–67
Alternaria, 166
Alzheimer's disease, 211–12, 251, 266
Ambylomma tick, 194
American College of Obstetricians and
    Gynecologists, 76
American Journal of Obstetrics and Gynecology,
    80
American Society of Microbiology, 93–94
American Type and Culture Collection, 244
amyotrophic lateral sclerosis (ALS), 210,
    251
animal bites, 53

animal husbandry, 134, 181, 253
anisakis worm, 144
Anopheles mosquitoes, 175, 181, 183, 186
anthrax, 52, 223, 243, 257
    bioterrorism and, 5, 236–39, 242, 246,
        249
antibiotics, 18, 20, 44, 69, 223–27, 254, 255
    how they work, 226–27
    multiheaded, 261–63
    overuse of, 72, 228–29
    for peptic ulcers, 203, 205
    Protective Response Strategy, 230-31
    resistance, 5, 13, 72, 222, 226–31, 251,
        254, 268
        germicides and, 232, 233, 234, 235
    for tuberculosis, 158–60, 161–62
antibodies, 42, 68, 186, 215
antigens, 67–68
antisense drugs, 259–60
Antiterrorism and Effective Death Penalty
    Act, 244
antitoxins, 69
Appert, Nicolas, 135
arenaviruses, 154, 155, 254
*Art of Preserving All Kinds of Animal and
    Vegetable Substances for Several
    Years, The.* (Appert), 135
asbestos, 162
Asnis, Dr. Debbie, 171
Aspergillus, 166
asthma, 163–69
    Protective Response Strategies for,
        168–69
Attila the Hun, 181
Aum Shinriyko, 242
autoimmune disorders, 69, 210–11

Ayurvedic medicine, 19–20

Babesia microti, 194
babesiosis, 193, 194
babies, breast-fed, 59, 106
Bacon, Roger, 21
bacteriophage therapy, 269–70
Bacteroides fragilis, 213
bad breath, 56–58, 85
balance, health and, 18
barber shops, germ threats in, 106
Barbour, Dr. A. G., 191
Barton, Clara, 119
Bartonella, 116, 194
bathing, 16, 23
    germs threats in the shower or bathtub,
        93, 106
B cells, 68
beauty salons, germ threats in, 106
bedding, *see* mattresses
Bellevue Hospital, 184
Bifidoacteria, 59
biopharming, 267
bioterrorism and biological weapons, 52,
        236–50
    agoterorrism, 246–48
    diagnosis problem, 241–42
    history of, 243–44
    likeliest biological weapons, 240
    Protective Response Strategy, 248–49
    scenario, 236–39
    vaccines to counteract, 241, 249
birds as virus hosts, 172–74, 177–79
Blaser, Dr. Martin J., 205, 206
blocking agents, 261
Block Island, 193
blood supply, prion disease and, 141–42
Boheme, La, 157
Bokkenheuser, Dr. Victor, 216
Borna disease virus (BDV), 211
Borrelia burgdorferi bacterium, 191–93, 195,
        196, 197
botanical drugs, 18, 268
botulism, 134, 243
bovine spongiform encephalopathy (BSE)
        (mad cow disease), 136–42, 147,
        256
brain cancer, 211
breast feeding, 59, 106

breast implants, silicone, 210–11
British Ministry of Agriculture, 139
bubonic plague, 20, 21
Bucharest, Romania, West Nile virus in, 174
Burgdorfer, Dr. Willie, 191
Burkitt's lymphoma, 207
Burnet, McFarlane, 182
Bush, George, 159
Bush, George W., 188

cachectin, *see* tumor necrosis factor (TNF)
calculus on and between teeth, 85–86
caliciviruses, 49
Campylobacter, 51, 95, 120, 143, 205
Campylobacter pylori, 202
cancer, 4, 5, 34, 164, 206–209, 254, 266
Candida albicans, 58, 268
carbon dioxide emissions, global warming
        and, 187–88
carcinogens, 162, 206
carpeting, 91–92, 169
carrion avoidance, gut morphology and,
        126–32, 133
cats, *see* pets
cat scratch fever, 194
Celera Genomics, 220
cell-mediated immune responses (CMI), 68
cells, 30–31
cellulose materials, *Stachybotrys* spores and
        water damage to, 162–64
Centers for Disease Control (CDC), 52, 106,
        111, 152, 153, 171, 174, 230, 233,
        249, 253, 274
    toxic shock syndrome and, 73, 75, 76, 78,
        80, 83
cervical cancer, 208–209
Chain, Ernest, 225
chemical weapons, 243, 244, 248
childbed (puerperal) fever, 24–25, 233
children, 106–11
    breast-fed babies, 59, 106
    immune system of, 15, 44, 52, 97, 107,
        147, 164
    Protective Response Strategies,
        108–109
    Stachybotrys spores and, 162–64
    toys, *see* toys
    vaccines, 107, 273
chimpanzees, 127–28, 129, 132, 154, 253–54

Chinese medicine, 4, 19, 20, 272–73
chlamydia, 48, 111, 112, 255
Chlamydia pneumoniae, 209
Chlamydia trachomatis, 208–209
chlorofluorocarbons (CFCs), 188
cholera, 27
cholesterol, 206, 258
chronic wasting disease (CWD), 140–41
ciprofloxacin, 222, 226, 237, 238
circumcision, HPV infection and, 208
Civilian Sanitary Commission, 119
Civil War, 118–20
Cladosporium, 166
cleanliness, *see* hygiene
Cleveland, Ohio, children dying from
         Stachybotrys spores in, 162–63
Clinton, Bill, 245
Clinton, George, 183
cloning, 264–65
Clostridium difficile, 258–59
Clostridium perfringens, 120, 129, 164
Cobra Event, The (Preston), 245
cockroaches, 166, 167, 169
cold, common, 54, 89, 98–103, 147, 245
    Protective Response Strategies for,
         101–102, 103
Colorado State University, 268
commensalistic relationships, 39
common-vehicle spread of germs, 51–52, 54,
         118–48, 275
condoms, 114, 208, 255
Convention on the Prohibition of the
         Development, Production, and
         Stockpiling of Biologic and Toxic
         Weapons and Their Destruction,
         1972, 244
cooking:
    history of, 132–33
    Preventive Response Strategies,
         143–46
    shellfish, 124, 144–45
coronoviruses, 99
cow's milk, babies fed, 59
Coxsackie virus (CSV), 218–19
Creutzfeldt-Jakob disease (CJD), 137,
         139–40, 211
Crick, Francis, 260
Crohn, Dr. B. B., *see* Crohn's disease
Crohn's disease, 212, 213

cross-reactive substances, 172
Cryptosporidium, 120, 122–24
Culex pipiens mosquito, 174, 175, 176, 177
cutlery, kitchen, 143
cutting boards, 143, 234
cyanobacteria (blue-green algae), 33, 34, 35,
         37
Cyclospora, 120, 124–25
cystic fibrosis, 265, 266
cytomegalovirus (CMV), 213
cytotoxic T cells, 68, 186
cytotoxins, 205

"Dateline," 90, 164
Davis, General G. W., 185
day care, 107, 166, 167
DDT, 185–86
DeArmond, Stephen, 137–38
decoys, 260–61
deer, 191–93, 196–97
DEET, 195
de Kruif, Paul, 8
dengue (breakbone) fever, 183
deodorants, 44, 60, 72
depression, 211
Dermatophagoides, 166
developing countries, 157, 167
    dirty injection equipment in, 256–57
    sanitation and contaminated water in,
         17, 27, 121, 148
d'Hérelle, Felix, 269
diabetes, 218–19, 265
diaper changing, 107
diarrhea and diarrheal disease, 107, 129, 167,
         213, 254, 258–59
    from contaminated food and water, 17,
         51, 119–21, 142
Dickens, Charles, 157
dietary laws, religious, 134
digitalis, 18
diphtheria, 257
Directly Observed Treatment Shortcourse
         (DOTS), 161
Dix, Dorothea, 119
DNA, 30–31, 260, 263
dogs, *see* pets
doorknobs, germs on, 97, 275, 276
DOTSPlus, 161
douches, 44, 60–61

doxycycline, 236, 238
draperies, 91, 92, 169
droplet spread of germs, 50
Duke University, 266
Dumas, Alexandre, *fils,* 156
duodenal ulcers, 204
dust mites, 166, 169
dysentery, 70, 119, 120

eastern equine encephalitis (EEE), 175, 178
Ebola fever, 5, 16, 114–15, 149, 153, 254
echovirus-7, 210
ectopic pregnancy, 209
eggs, raw, 144, 276
Ehrlich, Paul, 223, 224, 235, 243
Ehrlichia, 193
Ehrlichiosis, 193, 194
elderly, vulnerability to infectious germs, 15,
        44, 52, 145, 147, 164
    tuberculosis, 157, 161–62
Eliava Institute, 270
Eli Lilly, 269
encephalitis, 171–72, 210
endocrine system, 270–71
Entamoeba histolytica, 120
Enteritis necroticans, 130
Enterobacter sp., 14
enterococcus, 14, 222, 268
Enterococcus faecium, 228
enterotoxigenic B. fragilis (ETBF), 213, 214
Environmental Research Center,
        Smithsonian Institution, 144
epidemiology, 48
Epidermatophyton, 58
epiphytic growth, 86
Epstein-Barr, 207
erythema migrans (EM), 191
erythromycin, 226
Escherichia coli (E. coli), 14, 42, 44, 66, 70, 95,
        120, 145, 209, 222, 232, 268
    0157:H7, 70, 121–22, 145, 214, 270
    in vaginal vault, 60–61
esophageal cancer, 208
eukaryotic cells, 219–21
evolutionary digestion, 126–32, 133
expiration dates, food, 146

family, spreading infectious diseases within
        a, 97

fast food, 121–22
FBI, 246
fecal stasis, 216
fecal-to-oral route, contact spread of germs
        by, 50–51
Federal Emergency Management Agency
        (FEMA), 248
fish, raw, 144, 276
flaviviruses, 171, 172
Fleming, Alexander, 224, 225
Fleming, Ian, 247
Florey, Howard, 224–25
flu, 50, 89, 98–103, 120, 147
    Protective Response Strategies for,
        101–102, 103
fomite, defined, 89
Food and Drug Administration, 78, 80, 141
    Good Manufacturing Policy and, 85–86
    food contamination and food poisoning,
        50, 51–52, 64, 70, 121–22, 145,
        255–56
    diarrhea from, 17, 119–21, 142
    fast food, 121–22
    mad cow disease, 136–42, 147, 256
    produce, imported, 124–25, 135, 145
    protection from, 136, 230, 255
        Protective Response Strategies,
        143–48, 230, 231
    refrigerated foods, *Listeria monocytogenes*
        in, 125–26, 146
food lithe, 57–58
food-safety regulations, 148
foot and mouth disease (FMD), 247–48, 256
foot odor, 58–59
footwear, separating outdoor and indoor, 92
fossil fuels, global warming and, 187–88
Fracastoro, Girolamo, 4, 22, 25, 87
Freitas, Robert A., 272
fruits:
    cleaning, 145, 276
    imported, 124–25, 135, 145
Fukushima Medical University, 211
fungal molds and spores, 162–64, 166, 169

Galen, 20, 21
Garcia effect, 130–31
gargling, 57
General Accounting Office, 141
gene therapy, 217

genetic engineering, 26, 135–36, 244–45, 250, 263–65
genetic information in cells, 31
genetic therapy, 265–67
germicides, 147, 231, 232–35
germs, 277–78
    beginnings of the Earth and, 30–34
    classification of, 37–38
    coexisting with, 38–41
    functions of, 7–8, 34–36
    human beings as host of, see human flora
    number of known species, 36
    strategies for infiltrating host organisms, 69–71, 158
    transmission of, see transmission of germs
germ theory of disease, 212
    Fracastoro and, 4, 22, 25
    Koch and, 4, 26, 62–63, 63, 212
    Leeuwenhoek and, 4, 22, 24, 25
    Pasteur and, 25–26, 62, 119, 212
germ warfare, see bioterrorism and biological weapons
Giuliani, Rudolph, 179
glanders, 241–42
globalization, 5, 186
global warming, 152, 187–90
gonococcus, 69
gonorrhea, 48, 66, 69, 72, 112, 209
Good Manufacturing Policy, 85–86
Grassi, Giovanni, 183
Great Britain, 247, 248, 256
Greece, ancient, 19, 20, 156
greenhouse gases, 187–88
gut morphology, carrion avoidance and, 126–32, 133

Haas, Dr. Earl, 76, 77
Habitat for Humanity, 164
Haldane, J. B. S., 32
hand shaking, 48, 97, 102–103, 143, 276
hand washing, 15–16, 44, 102, 105, 136, 143, 230, 233–34, 255, 256, 276
    by hospital personnel, 5–6, 230, 234
    Protective Response Strategy, 27–29, 90, 95, 143, 231
    public education on, 6, 256
    in public rest rooms, 10–13, 29, 143
Hanna, Dr. Bruce, 78, 80–81
hantavirus disease, 152–53, 154, 155, 254

hantavirus pulmonary syndrome (HPS), 153
Harris, Elisha, 119
Hata, Sahachiro, 223
healers, ancient, 17–18
heart disease, 4, 5, 18, 206, 209, 222, 254
Heatley, Norman, 225
Helicobacter pylori, 166, 202–206, 258
hemolytic uremic syndrome (HUS), 121
hemophilia, 266
Hemophilus influenzae, 107
hemorrhagic fever with renal syndrome (HFRS), 153
Henson, Jim, 252, 253
hepatitis A, 50, 166
hepatitis B, 207, 256–57, 267
hepatitis C virus (HCV), 189–90, 207, 256
hepatocellular carcinoma, 207
herpes simplex viruses, 111–12, 207, 211–12, 266
HERV-W (retrovirus), 211
Hippocrates, 20, 221
history of hygiene and the understanding of disease, 17–27
    3000 B.C. to A.D. 500, 19–21
    A.D. 500 to A.D. 1500, 21
    A.D. 1500 to the present, 22–28
"History of Lactic Acid Fermentation, A," 25
HIV (human immunodeficiency virus), 4–5, 40, 48, 71, 112–14, 157–58, 189, 254, 261, 268, 274
    see also AIDS
home, germ threats in the, 90–97
Hooke, Robert, 4
hookworms, parasitic, 185
Horn, Floyd, 247
hospitals:
    antiseptic procedures in, 26, 119, 230, 232, 234
    infections picked up in, see nosocomial infections
    separate wards for specific diseases, 21
    showerheads, 151
Hospital Sketches (Alcott), 119
host environment, 64–65
    strategies of germs for infiltrating, 69–71
host organism, 48
Hughes, Howard, 3, 89
human flora, 37, 42–61, 275
    defined, 38

human flora (*cont.*)
  life cycle and development of, 47–48
  location and coexistence of humans with
      germs, 45–47, 66
  normal, 46, 65
  smell of, 56–61
  suitability of a site for, 43–44
Human Genome Project, 220, 274
human herpes virus-8 (HHV-8), 207
human papillomavirus (HPV), 111, 207–208,
      255
humidifiers, 169
humidity, 14, 19, 42, 53, 93, 125
Hussein, Sadam, 244
hyaluronic acid, 70
hygiene, 276
  food, *see* food contamination and food
      poisoning
  hand washing, *see* hand washing
  history of, *see* history of hygiene and the
      understanding of disease
  immigration and, 109–10
  public education about, 27, 256
  teaching children, 108–109

ileitis, *see* Crohn's disease
immigration, 109–10, 155, 168
immune modulation, 270–71
immune system, 15, 53, 67–71, 90, 223
immunosuppressed populations, 15, 40, 44,
      52, 123, 145, 147, 164
    Stachybotrys spores and, 163
    tuberculosis in, 157–58
    West Nile virus and, 174
  personality type and, 100
  prion infection and, 137–42
  tuberculosis and, 158
index case, 74
indigenous people, medicinal knowledge of,
      18
infectious diseases, 3–4, 22, 230, 253, 254,
      255
  within a household, 97
  transmission of, *see* transmission of germs
  see also specific diseases
infertility, 209
inflammatory bowel disorders, syndromes,
      and diseases (IBDs), 212–18
  diet and, 216, 217

holistic approach to treatment, 214,
      216–18
  multiple causes of, 212–13
influenza, *see* flu
Institute of Medicine, National Academy of
      Sciences, 76
insulin-dependent-diabetes mellitus (IDDM),
      218–19
interferon, 186
International Federation of Red Cross, 254
International Food Safety Council, 233–34
Iran, 160, 245
Iraq, 244, 245
irradiated food products, 148
isoniazide, 162
Israel, West Nile virus in, 174, 178, 179
IV drug uses, 114
Ixodes ricinus tick, 191–93, 194

Japan Unit 731, 243, 244
Jenner, Edward, 23–24, 273
Jisty, Benjamin, 273
Johne's disease, 213
Johns Hopkins University School of
      Medicine, 211
Joubert, Jules, 257
Journal of Anthropological Research, 127
juices, pasteurized, 145

Kaposi's sarcoma, 207
Kehm, Patricia, 73, 78
kissing, face, 48, 102–103, 276
kitchen, germ threats in the, 94–96, 234
  practicing basic food safety, 136
  Protective Response Strategies,
      143–48
Klebsiella pneumoniae, 15
Koch, Robert, 4, 26, 62–63, 212, 243
Koch's postulates, 62–63, 203
Kuczynski, Alex, 14
kuru, *see* Creutzfeldt-Jakob disease (CJD)

lactobacilli, 60, 61, 217, 258
Lactobacillus casei GG, 259
lactose intolerance, 215
Lady with Camellias, The (Dumas), 156
large intestine, digestive process in, 215–16
Larson, Jonathan, 157
Lassa fever, 154

laundry, 93–94, 169
Laveran, Alphonse, 183
Lederberg, Dr. Joshua, 245–46
Leeuwenhoek, Antoni van, 1, 4, 22, 24, 25
Legionella pneumophila, 70, 149–52
Legionnaire's disease, 149–52
  Protective Response Strategies,
    151–52
leukemia, 208
levofloxacin, 226
Libya, 244
linezolid, 235
Lister, Joseph, 26, 119, 232
Listeria, 70
Listeria monocytogenes, 125–26, 146
Lister Institute, 197
liver cancer, 207
loofahs, 93
Lou Gehrig's disease, 210, 251
Louis XIV, court of, 23
lung cancer, 208, 211
Lyme disease, 53, 190–93
  Protective Defense Strategies,
    195–97
Lymerix vaccine, 196
lymphocytes, 68
lymphomas, 207

macrophages, 158
mad cow disease, 136–42, 147, 256
malaria, 180–183, 185–87, 254
  Congress' campaign against, 185–86
  development of vaccine for, 186, 275
  history of, 181–183
  mosquitoes and, 20, 53, 54, 175, 180–83,
    185–87, 189, 253
Maloney, Carolyn B., 85
Marshall, Dr. Barry, 201–206
mattresses:
  dust mites in, 166
  Protective Response Strategies, 91,
    165
  reconditioned, 90–91, 164–65
Mayo Clinic, 90
measles, 21, 157
meats, 95
  carrion avoidance, gut morphology and,
    126–32, 133
  cooking, 132–33, 143–44, 276

imported, 147
mad cow disease, 136–42, 147
Memorial Sloan-Kettering Cancer Clinic,
  185
Mengele, Josef, 243
meningitis, 48, 171, 172, 210
menopause, 209
menstruation, 76–77, 209, 271
  smell associated with, 59–61
  tampon use, and toxic shock syndrome,
    76–85
Metchnikoff, Élie, 223, 224
methane, 36, 187, 188
methanotrophs, 36
Microbe Hunters, The (de Kruif), 8
microbiology, 26, 212
Microsporum, 58
Miller, Stanley, 32
money, spread of germs on, 104–105
monocytes, 68
mononucleosis, 110–11, 207
Montemagno, Carlo, 272
mosquitoes:
  dengue fever and, 183
  global warming and, 189
  malaria and, 20, 53, 54, 175, 180–83,
    185–87, 189, 253
  slash-and-burn agriculture and breeding
    grounds for, 181, 185–86, 253
  as vector for infectious germs, 172
  West Nile virus spread by, 172–80
  yellow fever and, 183–86
mousepox virus, 245
mouth:
  bad breath, 56–58, 85
  toothpastes, 85–86, 236
mouthwashes, 93, 232
mutualistic relationships, 39, 41, 42, 65, 69,
  277
Mycobacterium avium paratuberculosis (MAP),
  213, 214
Mycobacterium fortuitum, 106
Mycobacterium tuberculosis, 70, 157

Nader, Ralph, 84
nail salons, pedicure footbaths in, 106
nanotherapy, 271–72
nasopharynx, cancer of, 207
National Cancer Institute, 211

National Institutes of Health, 220
National Public Health Institute, Finland, 209
Nature, 217
nausea and vomiting, 131, 147
nervous system, 270–71
New England Journal of Medicine, 125–26
New York City, West Nile virus in, 170–80
New York City Department of Health, 171, 175, 178
New York Observer, 14, 16, 17
New York State, regulation of reconditioned mattresses in, 164–65
New York Times, 109, 248
New York University Medical Center, 80, 250
     Department of Clinical Microbiology, 171
New York University School of Medicine, 206
nitrates, 34–35
NK cells, 68
Nod 2 gene, 68–69, 217
nonoxynol-9, 61
North Korea, 245
Norwalk virus, 49, 267
nosocomial infections, 5–6, 13–14, 110, 230, 255, 256
nuclear energy, clean, 188
Nussensweig, Drs. Ruth and Victor, 186

obsessive compulsive disorder (OCD), 211
office, germ threats in the, 97–104
Ohio State University, Byrd Polar Research Center, 187
Oklahoma City, bombing of federal building in, 250
Olmsted, Frederick Law, 119
On Her Majesty's Secret Service (Fleming), 247
Oparin, A. I., 32
oral cancers, 208
organic compounds, 31–33
     breaking down of, into inorganic elements, 35–36
organic foods, 147
otitis media, 107
ozone layer, 33–34

Panama Canal, 184–85
pancreatic cancer, 208

Papanicolaou, George, 208
Pap smear, 208
para-aminobenzoic acid (PABA), 227
parasitic relationships, 39–40, 41, 42, 65
Pasteur, Louis, 4, 8, 25–26, 62, 63, 79, 119, 135, 199, 212, 223, 243, 257
Pasteur Institute, Paris, 189, 269
pasteurization, 26, 135, 145
Pasturella, 116
pathogenicity, 64, 66
Peloponnesian War, 74–75
pelvic inflammation, 209
penicillin, 4, 72, 73, 197, 222, 224–25, 226, 227, 235
Penicillium, 166
penile cancer, 208
peptic ulcers, 201–206, 258
peptide chains, 31, 33
perfumes, 44, 72
Pericles, 74
personality type and immune system, 100
pets:
     allergens, 166, 169
     Protective Response Strategy, 115–16
     toxoplasmosis transmitted by, 40, 116
     transmission of germs by, 115–17
phagocytes, 68, 70, 223
pharmaceutical companies, 18, 226
     peptic ulcer medications and, 203
photosynthesis, 30, 33, 34, 36, 188
Planned Parenthood, 111–12
plaque:
     arterial, 206
     dental, 85–86
Plasmodium parasite, 180–181, 182, 183, 186, 207
pneumonia, 15, 41, 50, 212–13, 257
pneumonic plague, 242–43
poliovirus, 266, 274
pollen, 166
polymicrobial infections, 55
polymyxins, 226
polyvinyl compounds, 162
population growth and urban sprawl, 155
poultry, 95, 143, 144
poverty:
     asthma and, 168
     Stachybotrys spores and, 164

pregnant women, vulnerability to infectious germs, 15, 40, 164
preschool programs, 107
preservation of food, methods for, 134–35
Preston, Richard, 245
prions, 137–42
probiotic therapy, 217, 257–59
Procter & Gamble, 77, 78, 80
produce, imported, 124–25, 135, 145
progesterone, 271
prokaryotic bacteria, 219–21
Protective Response Strategies:
    for allergies, 168–69
    for antibiotics, 230–31
    for arenavirus, 155
    for asthma, 168–69
    for bad breath, 57
    against bioterrorism, 248–49
    for carpeting and rugs, 92
    for children, 108–109
    defecating, proper way of wiping after, 50–51
    for food and water, 143–48, 230, 231
    for food odor, 58
    against global warming, 188
    for handling money, 105
    hand washing, 27–29, 90, 95, 143, 231
    for hantavirus, 155
    in the kitchen, 95
    for laundry, 94
    for Legionnaire's disease, 151–52
    for mattresses, 91, 165
    overarching, 275–76
    for pets, 115–16
    for sexually transmitted diseases, 114
    for *Stachybotrys* and other fungi, 163–64
    for tampon use, 83–84
    for telephone handsets, 96
    for ticks, 194–97
    toilet technique, 51 52, 92
    for tuberculosis, 160–61
    for TV remote controls, 96
    for water and food, 143–48, 230, 231
    for West Nile virus, 179–80
protein production, antisense drugs and, 259–60
proteins, 31
Proteus mirabilis, 210
protozoa, 38, 42

Pseudomonas aeruginosa, 15
Public Citizen Health Research Group, 141
Puccini, Giacomo, 157
pyruvates, 32

Q fever, 190
quorum sensing, 277

rabies vaccine, 26, 273–74
Ragir, Dr. Sonia, 127, 128, 130
Rajneeshi cult, 250
razors, 93
Reagan administration, 159
reconditioned mattresses, 90–91, 164–65
    Protective Response Strategy, 91, 165
refractory colitis, 213
refrigerated foods, *Listeria monocytogenes* in, 125–26, 146
Rely tampon, 77, 78, 80
Renaissance, 4, 22
Rent, 157
reservoir of infection, 48
rest rooms, hand washing in, 10–13, 29, 143
Rhazes, 21
rheumatic fever, 172
rhinoviruses, 49, 99
Rhone River delta, West Nile virus in, 174
Rickettsia, 193
Rifamate, 161
rifampicin, 162
RNA, 260, 263
Robert Debré Hospital, Paris, 217
Rockefeller, John D., 185
Rockefeller Foundation, 185
Rockefeller Institute, 185
Rocky Mountain spotted fever, 53, 190, 193, 220
rodents, disease spread by, 152–55, 169, 189
Roman Catholic Church, 21
Roman empire, 19, 20, 21, 181
Rosenberg, Dr. Martin, 127, 128, 130
Ross, Ronald, 183
rotavirus, 107, 167, 258
Royal Society of London, 22
rugs, 92
Russia, 160, 239, 245, 246

Safe Injection Global Network (SIGN), 257
Sagan, Carl, 31

Saint Louis encephalitis virus (SLE), 171–74
Salmonella, 51, 52, 53, 70, 95, 120, 143, 145,
  222, 228, 270
Salmonella typhosa, 222
Salvarsan 606, 224, 235
sanitation, 27, 155, 168
  in the developing world, 17, 27, 121, 148
  in Roman empire, 21
sarin, 242
satratoxins, 163
saunas, health club, 106
Schiotz, Jay, 257
schizophrenia, 211
Schumer, Charles, 250
scrapie, 137, 138
self-limiting infection, 16
Semmelweis, Ignaz, 24–25, 233
Senterfiet, Dr. Larry, 252
severe combined immune deficiency
  (SCID-X1), 265–66
sexual activity:
  ingestion of germs during, 51
  during menstruation, 60
sexually transmitted diseases (STDs), 48, 51,
  111–15, 207–209, 256
  Protective Response Strategies for,
  114
  safe sex, 255, 276
  see also specific diseases
shamans, 18
shellfish, 124, 144–45
Shigella bacteria, 53, 70, 120, 121, 269
Shinefield, Henry, 258
shower, germs threats in the, 93, 106
showerheads, cleaning, 151
"sick buildings," 162
sickle-cell anemia, 182, 265
sickle cell trait, 182
silicone breast implants, 210–11
skin cancer, 34
slave trade, malaria and, 181–182
sleeping sickness, 70–71
small intestine, digestive process in, 215
smallpox, 5, 21, 182, 245–46
  inoculation, 4, 20, 23–24, 272–73
  vaccine, 24
Smith, Theobold, 64
Smith equation, 64–65, 129–30
smoking, 163

soap, 19, 20, 93
  germicides, *see* germicides
Solanum nigrescens, 268
sorbitol, 215
Southern General Hospital, Glasgow, 266
Soviet Union, 269–70
Spanish-American War, yellow fever during,
  184
spas, health club, 106
spinal-cord injuries, 266
sponges, 93, 95
spontaneous generation theory of germs, 24,
  25–26
sports club and gyms, 58, 106
spread of germs, *see* transmission of germs
Stachybotrys spores, 162–64
staphylococcal infections, 147, 235, 257
Staphylococcus aureus, 14, 15, 41, 42, 43, 54,
  64, 70, 93, 145, 222, 226, 233,
  258, 268
  foot odor and, 59
  toxic shock syndrome and, 73, 74, 75, 81
Steere, Dr. Allen C., 190–91
steroids, 217
stomach viruses, 89, 120, 147
Strep group A, 15, 107, 172, 235, 252
Strep group B, 14, 107, 259
Streptococcus pneumoniae, 41, 42, 50, 54,
  69–70, 107, 222, 257
Streptococcus viridans, 14, 42
Streptococus pyogenes, 43, 252
streptomycin, 158, 225–26
stress, 100
"string of pearls" bacteria, 37–38
sulfa drugs, 227
sushi, 144, 276
Susruta, 20
symbiosis, 38–41
synercid, 235
syphilis, 22, 48, 112, 223, 224, 235

Tagamet, 202, 205
tampons and toxic shock syndrome, 76–85,
  86
  Protective Response Strategies,
  83–84
T cells, 68, 71, 204, 261
teeth:
  brushing, 57

teeth (*cont.*)
    toothpastes, 85–86, 232
telephone handsets, 96, 98
Terminalia chebula, 268
terrorism, germ warfare and, *see* bioterrorism
        and biological weapons
testosterone, 271
tetracycline, 226, 227
Texas Animal Health Commission, 141
Th-1 and Th-2 cytokines, 214–15
Third World, *see* developing countries
Thomas Jefferson University, 267
Thompson, Dr. Lonnie G., 187
Thucydides, 75
ticks, 53, 172, 190–94
    Protective Response Strategy for,
        194–97
Tierno, Dr. Philip M., Jr., toxic shock
        syndrome research of, 78–81
Tierno, Josephine, 78
Todd, Dr. James, 74
toilet:
    cleaning, 92
    closing lid before flushing, 92
    washing hands after using, *see* hand
        washing
toothbrushes, 93, 109
toothpastes, 85–86, 232
touch, transmission of germs by, 13, 27, 48,
        89, 90, 230, 256
towels, laundering, 93–94
toxic shock syndrome (TSS), 5, 8, 72–85,
        252
    biological factors, 80
    chemical factors, 79–80
    in history, 74–75
    lack of immunity after surviving, 81–82
    physical factor, 79
    symptoms of, 72–73
    tampon connection, 76–85, 86
toxoplasmosis, 40, 65–66, 116, 166
toys, 96, 108, 234–35
transmission of germs, 275
    airborne, 52–53, 54, 149–69, 275
    common-vehicle, 51–52, 54, 118–48, 275
    by contact, 13, 27, 48–51, 89–117, 90,
        107, 230, 253, 275
    environmental conditions and, 53–54, 64
    means of locomotion and, 54–55

vector-borne, 53, 54, 170–97, 275
Traviata, La, 156–57
Treponema pallidum, 223, 235
Trichophyton, 58
triclosan, 143, 232–33
Trypanosoma brucei, 70–71
TSS-like strep (streptococcal TSS), 252–53
tuberculosis, 5, 52–53, 64, 70, 72, 155–62,
        182, 243, 253, 254, 275
    the disease process, 157–58
    drugs to fight, 158–60, 161–62, 226
    failure to complete full course of treat-
        ment, results of, 159–60
    history of, 156–57
    immune system and, 70, 158
    multi-drug-resistance (MRTB), 159, 222,
        247–48
    Protective Response Strategies,
        160–61
Tufts University, 268
tularemia ("rabbit fever"), 190
tumor necrosis factor (TNF), 130, 158
TV remote control, 96
"20/20," 98, 105
typhoid fever, 95–96, 119, 193, 222, 243
Typhoid Mary, 95–96

Uganda, Ebola fever in, 114–15
ulcerative colitis, 212
ultraviolet (UV) light, 33, 34, 94
Union of Concerned Scientists, 229
U.S. Army Medical Research Institute of
        Infectious Diseases, 241–42
U.S. Congress, 185–86, 248
U.S. Department of Agriculture, 247, 248
U.S. Department of Defense, 240, 248
U.S. Geological Survey, National Wildlife
        Health Center, 178
U.S. Public Health Service, 185
University of Alabama, 266
University of Helsinmki, 209
University of Manchester, 181
University of Michigan, 217, 260
upholstery, 91, 169
urban sprawl, population growth and, 155
urease, 210
Urey, Harold, 32
urinary-tract infections (UTIs), 60, 209–10,
        222, 258

vaccines, 18, 68, 69, 254, 272–75
    against biological agents, 241, 249
    biopharming, 267
    childhood, 107, 273
    first effective, 63
    for Lyme disease, 196
    for malaria, research on, 186, 275
    new-wave, 272–75
    for pets, 115
    poliovirus, 274
    rabies, 26, 273–74
    smallpox, 24
    tuberculosis, 158
vacuum cleaners, 52, 92, 168–69
vaginal vault:
    acid environment of, 60, 77
    during menstruation, see menstruation
vancomycin, 222, 226
Vanderbilt University School of Medicne,
    205
vector-borne spread of germs, 53, 54,
    170–97, 275
vegetables:
    cleaning, 145, 276
    imported, 124–25, 135, 145
Verdi, Giuseppe, 156
Vibrio, 143, 144–45
viral infections, 99, 229
virulence of germs, 64, 66
VISA (vancomycin-intermediate-resistant
    S. aureaus), 222, 235
vitamin B6, 161

Wagner, Dr. Gorm, 80
Waksman, Selman, 225–26
water:
    in developing countries, 17, 27, 121, 148
    diarrhea from contaminated, 17, 51,
        119–21, 142
    early understanding of disease and clean,
        20
    municipal water supply:
    Cryptosporidium in, 122–24
    protection against Legionella in, 151
    Protective Response Strategies,
        143–48
    public works projects to provide clean, 27
    water damage in cellulose materials,
        Stachybotrys spores and, 162–64

Watson, James, 260
Wellcome Trust, 220
Wesley, John, 17
West Nile virus (WNV), 5, 16, 54, 170–80,
        185, 250, 254
    effects in its original habitat, 173–74
    scenario for spread of, 175–78
whirlpools, 106
white-footed mice, Lyme disease and,
        191–93
Whitewater Arroyo virus (WWA), 154
Whitman, Walt, 118
William, Montel, 136
wiping after defecating, Protective
        Response Strategy for, 50–51
Withering, William, 18
workplace, germ threats in the, 97–104, 276
World Health Organization, 27, 53, 158, 160,
        206
World Meteorological Organization, 33–34
World Trade Center bombing of 1993, 250
World War I, 243
World War II, 243–44
Wright, Dr. Robin, 202

xylose, 215

yellow fever, 183–85
Yeltsin, Boris, 239
Yersinia pestis, 242–43

Zantac, 202, 205